COLOR ATLAS
OF
Venereology

ANTHONY WISDOM
MB, BS

*Consultant Venereologist to Oldchurch, South Essex
& Thames Groups of Hospitals*

YEAR BOOK MEDICAL PUBLISHERS, INC.
35 E. WACKER DRIVE–CHICAGO

Distributed in Continental North, South and Central America,
Hawaii, Puerto Rico, and the Philippines by
Year Book Medical Publishers, Inc.
By arrangement with Wolfe Medical Publications Ltd.
Printed by Smeets-Weert, Holland.
Library of Congress Catalog Card Number: 73-76996
International Standard Book Number: 0-8151-9329-7

Series Editor, G Barry Carruthers MD (Lond)

Books in this series already published
Color Atlas of General Pathology
Color Atlas of Oro-Facial Diseases
Color Atlas of Ophthalmological Diagnosis
Color Atlas of Renal Diseases
Color Atlas of Venereology
Color Atlas of Dermatology
Color Atlas of Infectious Diseases
Color Atlas of Ear, Nose & Throat Diagnosis
Color Atlas of Rheumatology
Color Atlas of Microbiology
Color Atlas of Forensic Pathology
Color Atlas of Pediatrics
Color Atlas of Histology
Color Atlas of General Surgical Diagnosis
Color Atlas of Physical Signs in General Medicine
Color Atlas of Tropical Medicine and Parasitology
Color Atlas of Human Anatomy
Color Atlas of Cardiac Pathology
Color Atlas of Histological Staining Techniques
Atlas of Cardiology, ECG's and Chest X-Rays

Some further titles now in preparation
Color Atlas of Neuropathology
Color Atlas of Oral Anatomy
Color Atlas of Oral Medicine
Color Atlas of Gynecological Surgery (6 volumes)
Color Atlas of Tumors of the Eye
Color Atlas of Liver Diseases
Color Atlas of Periodontology
Color Atlas of Diabetes Mellitus

To the Association of Technicians in Venereology

ACKNOWLEDGEMENTS

Illustrating this Atlas would have been impossible without the help and kindness of many friends and colleagues who have most generously allowed publication of photographs from their collections. I would particularly like to thank (and I apologise for any omissions) the following :
Dr Suzanne Alexander, Dr June Almeida, Dr J A Armstrong, Dr R D Catterall, Dr G Csonka, Dr E M C Dunlop, Dr A Grimble, Dr J A H Hancock, Dr M J Hare, Mr I A Harper, Dr R N Herson, Dr J J Jefferiss, Mr L Kay, Dr G M Levene, Dr C S Nicol, Dr J K Oates, Dr J D Oriole, Dr G Rohatiner, Mr H Thompson, Dr D Vollum, Dr A E Wilkinson and Dr R R Willcox; Drs Borchardt and Hoke of San Francisco, Dr Vernal G Cave of New York, Dr W C Duncan of Baylor College, Professor Paolo Nazarro of Rome, and Dr T Guthe of WHO; the photographic departments of Guy's Hospital, the London Hospital, the Middlesex Hospital, the Westminster Hospital, Dr Cardew and the Photographic Department of St Mary's Hospital, Mr E A Shepherd of Oldchurch Hospital, the Trustees of the Wellcome Institute, Messrs May and Baker Ltd and the Venereal Disease Research Unit Center for Disease Control, Atlanta, Georgia, USA. I am also very grateful for the constant help given by the nursing staffs at my clinics, and for the forbearance and encouragement of the editor of the Wolfe Medical Atlases, Dr G Barry Carruthers.

CONTENTS

INTRODUCTION

Throughout the world, medical and public health authorities are concerned with the frightening increase in the numbers of patients with sexually transmitted diseases. Clearly, one of the most important factors in control is effective treatment, which *must* be preceded by accurate diagnosis. The increasing complexity of modern medicine is reflected in the medical school curriculum and it is difficult for the student to give more than a short time to the problems associated with venereal diseases: problems not only of diagnosis and treatment but also the very real psychological problems which are so common in patients, their partners and families when venereal disease is suspected or found.

To put it in a nutshell, the reason for attendance at a venereal disease clinic falls into one of three categories. These categories are :
1 Development of symptoms suggestive to the patient, or to a referring practitioner, of sexually transmitted disease.
2 After exposure to possible infection to exclude disease.
3 As a contact of a sexual partner found to have a transmissible condition.

The final assessment after evaluation of history, examination and results of investigations will also fall into one of three categories:
1 Sexually transmitted disease present, often with findings of genital or extra-genital lesions: treatment is usually required.
2 Genital lesions present which are manifestations of conditions not primarily associated with sexual transmission: local or other forms of treatment may be required.
3 No genital disease or lesion present: this category will include many anxious or phobic patients who may require other forms of treatment or referral.

This Atlas is designed for students and non-specialist medical practitioners and others interested in the subject to give an outline of techniques of examination and presentations of sexually transmitted and other diseases likely to be encountered in clinical practice and to illustrate some of the lesions in these conditions. Throughout, I have tried to emphasise the more common manifestations. I am well aware that many rare conditions are not illustrated and that I have omitted much 'small print'. For fuller information I would recommend readers to consult some of the excellent textbooks and monographs that are available. I have also omitted references and methods of treatment: the former because they are unnecessary in this type of book and the latter because therapeutic practice changes rapidly.
    I hope that this Atlas will facilitate the recognition of the conditions depicted and therefore be instrumental in the achievement of earlier

diagnosis and the institution of effective treatment. Finally, I emphasise that 'treatment' in the context of sexually transmitted conditions does not only mean chemotherapy or other physical measures but must include sympathetic understanding and reassurance of the patient who is so often acutely anxious about his or her condition.

Anthony Wisdom, London, February 1973

AETIOLOGY OF COMMON PRESENTATIONS

INTRODUCTION

The object of a consultation at a venereological clinic is for a conclusion to be made, either the diagnosis or exclusion of disease. This objective can only be achieved by appraisal of the history, and by examination and investigation. Patients present for examination either with symptoms of abnormalities of structure or function, or without symptoms when anxious, or as contacts of others already found to have transmissible conditions. Structural abnormalities may be due to disease or to anatomical anomaly: other symptoms may be due to pathological or physiological causes; asymptomatic attenders include some without disease but others (including many women and male passive homosexuals) in whom disease can only be found by examination. In this section the evaluation of the history, observed abnormalities and the more common presenting symptoms is outlined, but it is most important to remember that the large group of totally asymptomatic patients equally require careful examination and assessment.

History Points that should always be noted when taking the history are listed below:

Current symptoms (if any) and duration. Direct questioning is often helpful.

Current and recent therapy (prescribed or self-administered).

Current general health.

Previous sexually transmitted or other genito-urinary disease, general medical history and family history. In patients originating in areas where treponemal disease is endemic, specific enquiry should be made of an history of these conditions.

Timing of occurrence and type (e.g. marital, casual, homosexual) of recent sexual contacts and any prophylaxis used.

Presence or absence of genito-urinary symptoms in sexual partner.

(In women) obstetric, gynaecological and menstrual history and contraceptive methods.

The history and examination should always be conducted in a systematic manner: the symptoms and findings will determine further investigations to be undertaken. In the interests of clarity it is most important that the words and phrases used are understood by both the interviewer and interviewee: medical personnel must always remember that some words and phrases that have a specific connotation to themselves may be used in a quite different sense by patients. Furthermore, many patients will be ignorant of the correct terminology used to describe sexual actions or anatomical parts: it is prudish to refrain from using the vernacular in situations where this would appear to be helpful.

Anatomical anomalies

Variance from the 'normal' is a common reason for attendance or an anomaly may be found fortuitously. Patients may attend for examination because of a belief that the anomaly, often recently noticed, is a sign of sexually transmitted disease. Recognition is usually straightforward but some conditions, such as phimosis or hypospadias, appear to predispose towards infection and are found more frequently in venereological practice than in the general population. The illustrations in this sub-section are a selection of anomalies observed in clinical practice.

1 Hypospadias. The urethral opening is situated on the ventral surface of the penis: the lesion may occur anywhere between the glans and the perineum, and is most frequent distally. The glans may show a groove in the mid-line. (N.B. the penis is rotated in the photograph.)
2 Hypospadias. The bougie demonstrates the aberrant ventral orifice with canalisation of the glans.
3 Hypospadias. Note oedematous burrows of scabies.

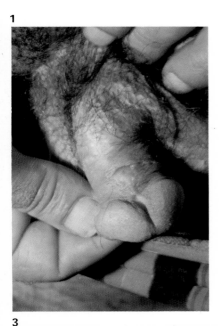

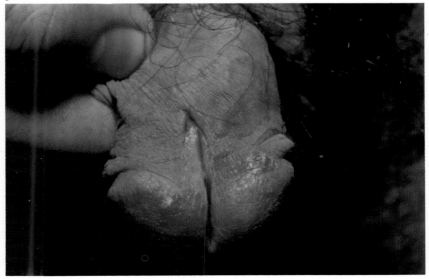

4 Median raphe duct. The urethra was normal. The median raphe has remained canalised, causing a short blind sinus.

5 Para-urethral duct. A common anomaly: usually the duct is shallow and does not communicate with the urethra; rarely, multiple ducts occur.

6 Paired median raphe ducts. A rare anomaly, resulting in this case in the formation of two short tunnels (demonstrated by the bougies).

7 Dorsal duct. A very rare anomaly (I have never seen another). This patient presented with gonococcal infection of the duct and the urethra and defaulted before investigation of the course of the duct could be undertaken.

8 Oedema of the glans penis. The glans is indented by the pattern of mesh underwear, the patient had non-gonococcal urethritis.

9 Mucocoele. Firm translucent thickening of lymph channels proximal to the coronal sulcus. The lesion is due to blockage of the lymphatics which may follow trauma or infection, but may also occur spontaneously. It is of no significance but may arouse anxiety. In this case the lesion was the sole reason for attendance. This condition is also known as *sclerosing lymphangiitis*.

4

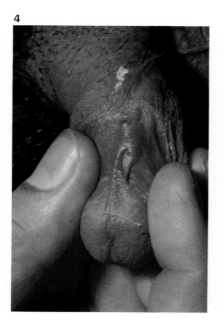

5

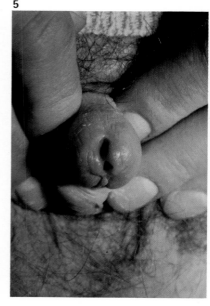

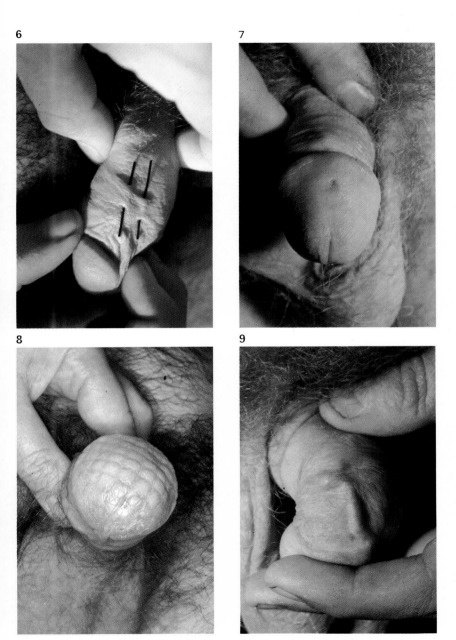

10 Para-urethral naevus. Naevi may occur anywhere on the genitalia and occasionally cause suspicion of venereal disease.

11 Varicosity of fraenal veins.

12 Phimosis. The opening is too small to allow retraction of the prepuce. The condition may be present from birth or may follow contraction resulting from trauma, infections or balanitis xerotica obliterans (see p. 285).

13 Paraphimosis. The glans has passed through a contracted preputial opening: the contracted band obstructs drainage causing gross oedema (here confined to the prepuce) distally.

14 Keloid. In the past this negro patient had balano-posthitis with adhesions and keloid scars have developed on the glans and prepuce.

15 Preputial adhesion and balanitis. Partial adhesion of the prepuce to the corona has resulted in difficulty in achieving adequate hygiene.

10

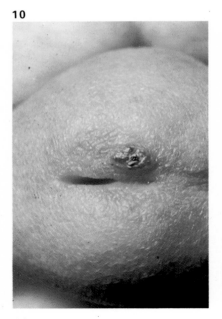

11

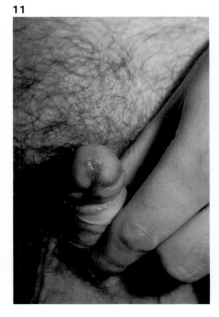

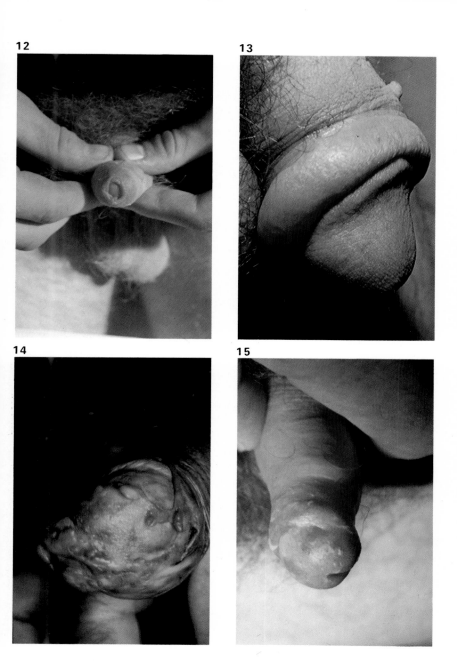

16 Depigmentation. This depigmentation followed scalding of the genitalia in childhood (Cf. fig. 19.)

17 'Pimples'. A surprisingly frequent reason for attendance in adolescents, who became anxious when such (physiologically) small sebaceous glands and hair follicles on the shaft of the penis are first noticed.

18 Striae. This patient attended convinced that he had 'secondary syphilis' after reading an ancient textbook. No physical disease was found; he had gained considerable weight in the recent past.

19 Vitiligo. Affecting lower abdomen and penis: in this patient the lesions were a source of anxiety.

20 Varicocele. A fairly common reason for attendance in adolescence. The characteristic 'bag of worms' sensation on palpation is diagnostic.

21 Angiokeratoma of the scrotum. The small blue lesions are a degenerative change and usually appear in middle-age or later: conscience may cause ascription to venereal disease. Similar lesions may be found on the vulva.

16

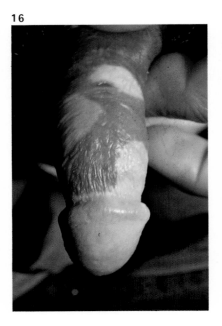

17

18

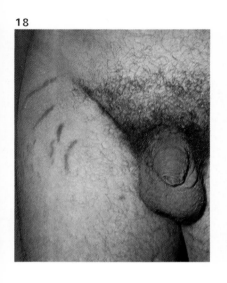

19

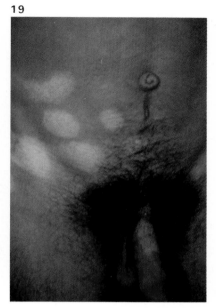

20

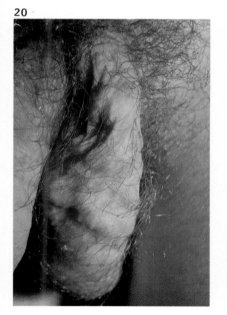

21

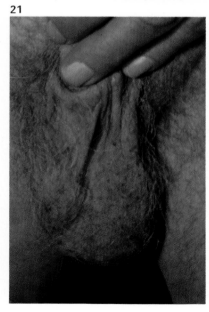

22 Multiple sebaceous cysts of the scrotum (steatocystoma multiplex). Small sebaceous cysts are seen commonly and may become infected: multiple lesions such as these are rare.

23 Hydrocele. The tense, transilluminable scrotal swelling is usually chronic but occasionally causes attendance in anxious individuals.

24 & 25 Smegma. Lack of hygiene may result in gross accumulation of smegma which may provoke anxiety, or may sometimes cause balano-posthitis.

26 Angiokeratoma of the scrotum, *(Fordyce)*.

27 Anal tags. Usually related to piles or a fissure, they may cause anxiety in male homosexuals.

22

23

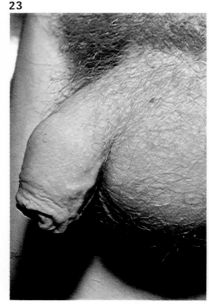

24

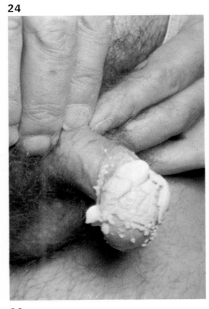

25

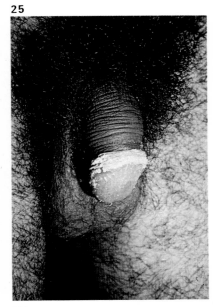

26

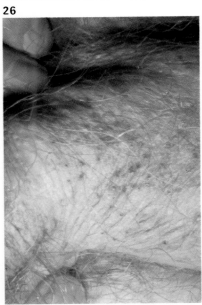

27

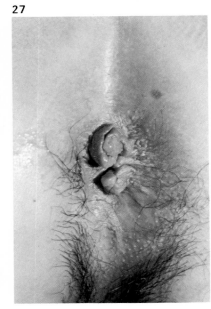

28 Vulval cysts. These cysts are usually inclusion cysts following trauma, especially during childbirth. Similar cysts occur on the vaginal wall and may be due to mesonephric remnants (Gartner's cysts). The cysts are usually noticed fortuitously but occasionally become infected or cause dyspareunia.
29 Cysts of Skene's ducts. Such lesions are usually asymptomatic. Urethrocele may present a similar appearance.
30 Vulval varix. Often prominent in pregnancy.
31 Pedunculated fibroma of labium majus.
32 Hypertrophied hymenal remnant (tag). These tags frequently become oedematous during intercourse: hymenectomy is often necessary.

28

29

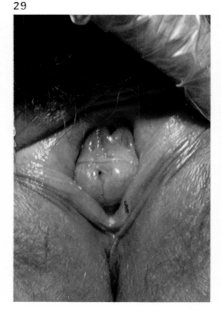

30

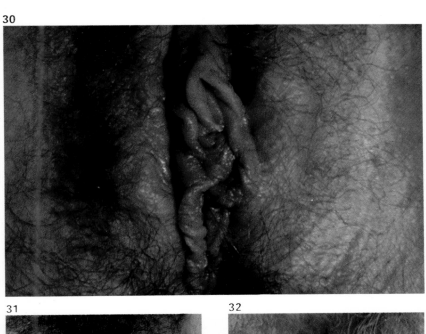

31

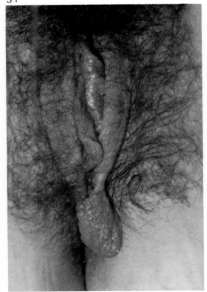

32

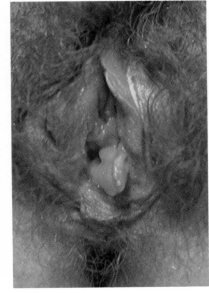

33 Capillary haemangiomata of vulva.
34 Bartholin's cyst. Small asymptomatic Bartholin's cysts are due to obstruction of the duct. No treatment is required unless they interfere with intercourse or become infected.
35 Pinhole os and tiny polyp.
36 Cervical polyp.
37 Cervical erosion.
38 Cervix showing IUCD threads.

33

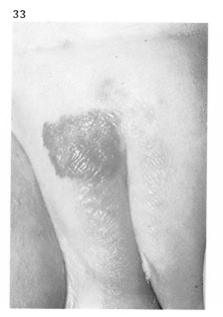

34

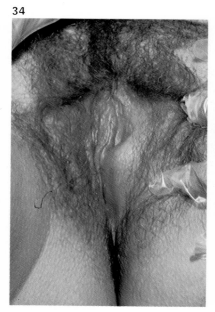

35

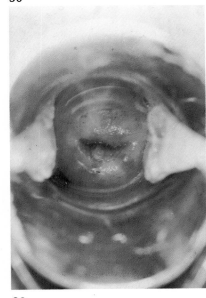

36

37

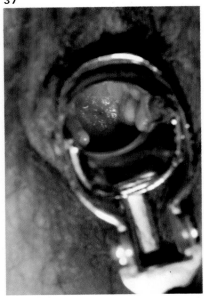

38

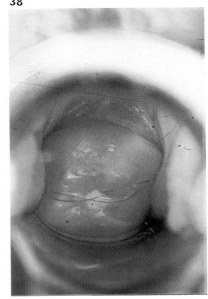

Genital ulcers

The presenting complaint is usually an *ulcer* or *sore* which has appeared on or adjacent to the genitalia. Male patients may attend with symptoms of balano-posthitis or preputial oedema: female patients may attend complaining of vulval swelling (oedema): examination reveals underlying ulceration. Male passive homosexuals (and some females) may develop lesions at the anus or on the buttock. Occasionally the initial complaint is of enlarged and/or tender inguinal lymph glands (see p. 26): examination shows active or healed genital ulceration. In some cases lesions on remote parts of the body indicate the need for genital examination which may reveal hitherto unnoticed ulceration. The term *sore* is also used by patients with conditions such as genital warts, herpes, furunculosis or balano-posthitis or it may refer to dysuria or other urinary symptoms.

The importance of a comprehensive history cannot be over-emphasised: some patients (*e.g.* homosexuals) are too embarrassed to be truthful on initial attendance and the true history may only emerge when the confidence of the patient has been gained. In conditions such as malignant ulceration, recollection of past (mis) deeds may cause the patient to fear sexually transmitted disease and to consult a venereologist in the first instance.

The list of causes opposite is not comprehensive: the list shows the more frequent and important conditions likely to be encountered in venereological practice.

Aetiology of genital ulcers

Group	Cause	Notes
Infections	Syphilis: primary, secondary, rarely tertiary	
	Lymphogranuloma venereum	Principally endemic in tropical and sub-tropical areas
	Chancroid (soft sore, ulcus molle)	Principally occurs in tropical areas
	Herpes	Recurrent. The most common cause of genital ulceration
	Pyogenic	
Trauma	Mechanical damage	May be self-inflicted or due to injudicious prophylaxis
	Chemical damage	
Neoplastic	Carcinoma	
	Pre-malignant conditions e.g. erythroplasia of Queyrat; Paget's disease	
Allergic	Fixed drug eruptions	History often diagnostic
	Generalised reactions N.B. Stevens–Johnson syndrome	
Secondary	In parasitic infestation e.g. scabies, pediculosis	Infestation usually evident at other sites
	Irritant dermatoses	Usually diagnostic evidence elsewhere
Unknown	Behçet's disease	Usually concomitant lesions of other sites (e.g. mouth, eyes) are present

Lumps in the groin

Lumps in the groin encountered in practice are nearly always of pathological significance but the cause is often unrelated to sexually transmitted disease or other anogenital conditions. The mass may have been present for some time before attention is sought: conditions such as ectopic location of the testis, hydrocele of the spermatic cord in males and of the canal of Nuck in females and inguinal hernia (direct or indirect) may present in this way. Lesions found below the inguinal ligament that may present in a similar manner include femoral hernia, saphenous varix and arterial aneurysm.

The more significant masses in the groin are due to *enlargement of the lymphatic glands (bubo),* which are normally impalpable although they may sometimes be detected in thin individuals. Glandular enlargement may be secondary to pyogenic anogenital conditions (often infected parasitic lesions) or to lesions of the leg or abdominal wall: moderate enlargement, often bilateral, and tenderness of the glands is found. Patients with urethritis, balano-posthitis, vulvitis or vaginitis (especially in cases of herpes) are occasionally found to have moderate tender glandular enlargement in association with other symptoms: the glandular symptoms are unlikely to be the presenting feature.

Glandular enlargement in *primary syphilis* is found in about 50 per cent of cases, and may also be found as part of the generalised lymphadenopathy in secondary syphilis. Examination of the syphilitic buboes shows the glands to be firm and rubbery to palpation and moderately enlarged ; aggregation and cutaneous involvement does not occur. Moderate tenderness may be experienced but often the enlarged glands are painless.

In *lymphogranuloma venereum* the advent of the bubo is likely to be the first sign of infection, especially in male patients. The bubo may be unilateral or bilateral and is usually painful, irregular and hard, with the glands matted together. Multilocular areas of softening and multiple sinus formation may occur. The skin overlying the bubo is often erythematous and thickened. Characteristically the glands both above and below the inguinal ligament are involved so that the lesion appears to be traversed by a groove.

In *chancroid* the bubo is similar to that found in lymphogranuloma venereum but the genital primary lesion is usually present. When softening occurs this is unilocular and when a sinus is present this is usually single and associated with adjacent cellulitis.

In *granuloma inguinale* the mass that is often found in the groin arises from subcutaneous spread of the infection. Abscess and sinus formation may occur: the epithelium adjacent to the sinus shows the characteristic beefy-red granulation tissue.

It is important to remember that enlarged glands may also be due to conditions such as glandular fever or to malignant disease. Reticuloses may first appear in the groin or the glands may be the site of secondary growth from anogenital or remote primary malignant lesions.

Aetiology of lumps in the groin

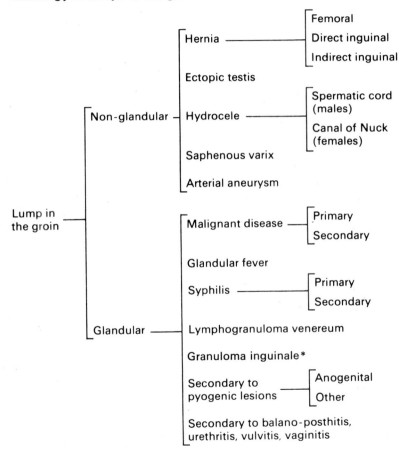

*The lump is a pseudo bubo rather than a true bubo (see page 170).

Anal and perianal region

Symptoms originating in this area are relatively uncommon in vereological practice but may occur in male passive homosexual, some female and, occasionally, in male heterosexual patients. Asymptomatic infections are much more frequent. The commonest symptom is *itching (pruritus ani)*: it is most important to remember that this symptom is frequently related to stress or anxiety and that patients may associate the complaint with past, guilt-provoking, sexual behaviour. Pruritus ani due to organic causes may originate from non-venereal conditions such as haemorrhoids, anal fissure or dermatoses; from venereal conditions such as warts, herpes or other ulceration; or from rectal discharge secondary to conditions such as proctitis or threadworm infestation. Symptomatic candidiasis associated with antibiotic therapy often begins with perianal irritation which may later spread anteriorly. In female patients perianal symptoms may be due to posterior spread from inflammatory vulval conditions such as candidiasis or trichomoniasis. Pain in the anal or perianal region may result from excoriation of irritant lesions or to conditions such as anal fissure. Complaint of *rash* in the area may refer to warts or to mycotic or herpetic infection. Tenesmus and constipation are occasionally associated with proctitis, but more commonly with anal fissure or prolapsed piles. *Discharge* from the rectum may be noticed as a sensation of anal dampness or seepage, or the patient may become aware of mucoid or purulent discharge on the faeces: proctitis is the usual cause. Anxious patients with anal tags or prolapse may occasionally consult a venereologist in the first instance.

AETIOLOGY OF COMMON PRESENTATIONS

Anal and perianal symptoms

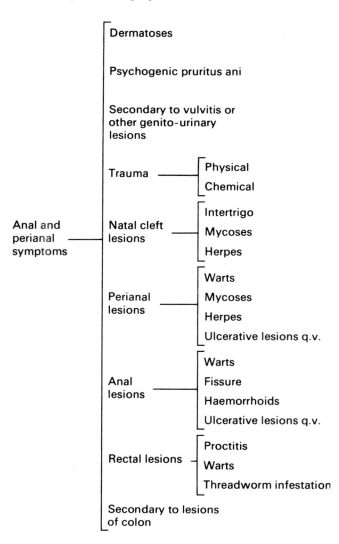

MALE PATIENTS

Introduction

The majority of male patients (apart from passive homosexuals) with
non-ulcerative lesions of the penis and adjacent areas have symptoms
and/or abnormal signs at the time of attendance; asymptomatic presentations (contacts or anxious individuals) are less common. In this section
the more frequent presenting symptoms (apart from those mentioned
previously) are discussed. These symptoms most commonly originate from
the prepuce, glans penis and anterior urethra. Less frequently, symptoms
may originate from lesions of the scrotum or its contents or more proximal
parts of the urinary tract and related structures. The evaluation of genito-
urinary symptoms in male patients is generally easier than in female patients
because the symptoms and their causes are much more likely to be
anatomically specific. Lumps in the groin have already been discussed on
page 26 and anogenital symptoms on page 28.

Discharge

Complaint of *discharge* usually refers to secretion observed at the end of the
penis and is the commonest symptom (but seldom the only one) in practice.
Discharge may originate from the urethra or from beneath the prepuce:
rarely, the origin may be from another orifice such as Tyson's gland or a
median raphe duct. Discharge may be due to physiological causes (*e.g.*
prostatorrhoea, nocturnal emission), local pathological causes (e.g. balano-
posthitis, urethritis, trauma) or to lesions of the upper urinary tract: sometimes the symptoms may originate from a combination of causes. The
complaint of urethral discharge or dampness may be made by anxious
patients who have become over-aware of genital sensation. Some patients
complain of intermittent discharge: patients with mild urethritis may only
notice the discharge in the morning or after urine has been retained for some
hours, and patients with prostatorrhoea may only notice the discharge after
defecation. Patients with crystalluria may complain of 'chalky' discharge
which occurs at the end of urination. Occasionally discharge may only
be evident by stains appearing on the underwear (although this symptom
is usually due to over-anxiety) or by a crust forming at the urinary meatus.
  The history and examination will usually differentiate the anatomical
origin of the discharge; occasionally differentiation is impossible because
the patient has an unretractable prepuce (phimosis). Profuse, thick,
purulent urethral discharge is likely to be due to gonorrhoea; scanty, thin
discharge is likely to be due to non-gonococcal urethritis; sub-preputial
discharge is likely to be due to balano-posthitis, often caused by elementary

lack of hygiene but, if lumpy and irritant, likely to be due to candidiasis. The character of the discharge is a diagnostic clue but bacteriological examination is essential.

Aetiology of discharge in males

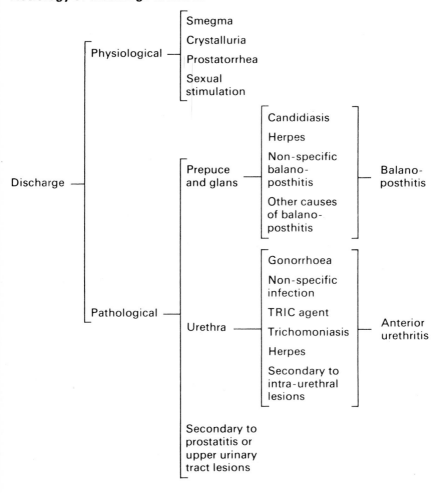

Urinary symptoms

Urinary symptoms are extremely variable and may originate from inflammatory conditions of the prepuce, meatus, anterior urethra or more proximal parts of the urinary tract, from crystalluria, or, in anxious patients, may occur without objective abnormal signs. The commonest symptoms are *pain* or *irritation* with urination; these are usually associated with meatitis or urethritis but can occur when urine comes in contact with an inflamed or fissured prepuce. Severe pain and scalding (traditionally 'like razor blades') may even lead to retention: this is most often associated with gonococcal urethritis (see p. 186). Milder urethritis is more often associated with complaint of irritation, initially located at the meatus but later passing proximally to the fossa navicularis and urethra. Usually pain and irritation are exacerbated by urination but are sometimes present at other times: symptoms are often most marked after urine has been retained for some hours or overnight. Supra-pubic pain indicates cystitis or upper urinary tract involvement; perineal pain with urination sometimes occurs in patients with prostatitis.

Urgency and/or *frequency* of urination are other common complaints which may occur alone (particularly in anxious individuals) or in association with other symptoms. This symptom may be due to meatitis or urethritis or may be due to involvement of the proximal urinary tract or prostate.

The early stages of urethritis or meatitis sometimes cause the edges of the urinary meatus to adhere so that the patient may notice difficulty in initiating urination, or that the urinary stream is split into several divergent pathways causing *spraying;* in patients who have experience of infection in the past this may be the first indication of trouble and the reason for attendance.

Haematuria occasionally occurs : it is most frequently associated with cystitis, particularly with the condition known as acute haemorrhagic cystitis (see p. 204).

Obstructive symptoms (difficulty in commencing urination, poor stream, post-urination dribbling) are infrequent but if present may indicate involvement of the prostate or development of urethral stricture. Anxious patients are prone to complain of urinary dribbling or stained underwear following treatment but in the majority of these individuals no objective signs of persistent disease can be found.

Aetiology of urinary symptoms in males

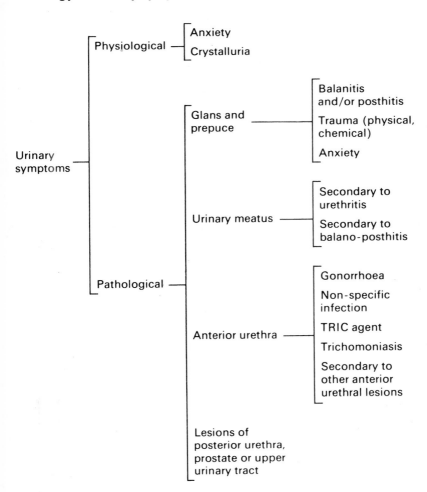

Scrotal symptoms

Symptoms originating from the scrotal area are relatively infrequent, and in many cases are not of pathological significance. Symptoms may originate from the surface or from within the scrotum; the cause may be due to local lesions or secondary to lesions elsewhere. Anxiety and attendance may be occasioned by the patient (often adolescent) becoming aware of a scrotal *rash* which on examination is found to be the normal or slightly prominent hair follicles and sebaceous glands, or by the discovery of a varicocele. Anxiety may be aroused by physiological activity of the dartos muscle: the resultant scrotal and testicular movement is thought to be abnormal.

The commonest scrotal symptom is complaint of a *rash* which is often irritant. The rash may be localised or involve all the scrotum and may also involve adjacent areas of the thigh, groin, or penis. Localised irritating rashes are most commonly due to folliculitis or scabies, while generalised irritating rashes are commonly due to fungal infections or other dermatoses. Scrotal sebaceous cysts (fig. 22) may enlarge and occasionally become infected. Other scrotal skin lesions are uncommon.

Symptoms arising inside the scrotum may be due to conditions affecting the spermatic cord, the testicle and its coverings or the epididymis. The only common lesion of the spermatic cord likely to cause complaint is a cyst or spermatocele, but the cord may be tender or thickened in patients with epididymitis. *Painless swellings* within the scrotum are usually found to be hydroceles but the symptom may rarely be associated with other conditions such as tuberculosis, neoplastic disease or late syphilis. In tropical areas enormous scrotal swellings may be due to lymphogranuloma venereum (see p. 176) or filariasis but presentation at venereological clinics is unlikely. *Painful lesions* may originate from the epididymis or testis; differentiation is impossible on history and may be difficult on examination. The commonest painful lesion is epididymitis, usually secondary to urethritis or other urinary tract infection; inflammatory hydrocele is a common associated finding. Other causes of painful lesions are torsion of the testis (which is often a recurring condition) and mumps orchitis. Rarely, the development of epididymitis is the first symptom of gonococcal or other urethritis.

Aetiology of scrotal symptoms

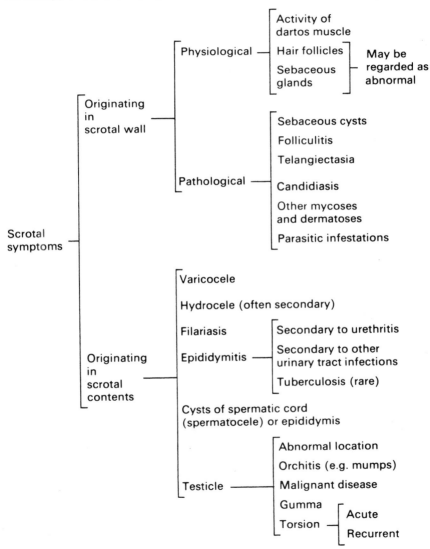

Preputial symptoms

Preputial symptoms may be due to local conditions affecting the glans penis and prepuce (balano-posthitis, see p. 244) or may be secondary to conditions originating elsewhere. Many patients with preputial symptoms have *phimosis* (see fig. 12), a condition which makes adequate hygiene and examination difficult to achieve. The commonest complaints appearing to originate from the prepuce are those of discharge or urinary disorder; these have already been discussed (p. 30 and p. 32). Other common preputial symptoms are those designated *rash, cut* and *swelling.*

The term *rash* is frequently used to indicate erythema of the prepuce (and usually of the underlying glans) which may be due to any of the causes of balano-posthitis or may refer to conditions such as genital warts or herpes. Irritant erythema is often due to candidiasis and may be due either to fungal infection or hypersensitivity. Localised irritating papules may be due to scabies (see p. 294) or to irritant dermatoses such as lichen planus.

The term *cut* usually refers to small fissures appearing at the preputial margin, often following intercourse, or it may be used in respect of ulcerative conditions. The symptom tends to be recurrent as the vicious circle of tissue damage → scarring → increased liability to further damage occurs. Fissuring is common when phimosis is present and is often exacerbated by concomitant infections. It is frequently seen in patients with balanitis xerotica obliterans (BXO) (see p. 285) and candidiasis (see p. 229).

Swelling (oedema) of the prepuce is frequently associated with trauma (often sexual activity) but may also occur with balano-posthitis or urethritis and is particularly marked if a phimotic prepuce is retracted and becomes caught behind the glans penis (paraphimosis, see fig. 13); oedema is usually more marked on the ventral surface.

Pale areas occurring at the preputial margin (often associated with phimosis) or patchily on other parts of the prepuce, penile shaft or the glans penis may occasionally cause attendance. In most cases these lesions are due to BXO but (particularly in coloured patients) dyschromia may follow trauma or inflammatory conditions, or may be due to a congenital cause such as vitiligo.

Skin symptoms

Cutaneous symptoms originating from the prepuce, perineum and scrotum have already been discussed in preceding paragraphs but symptoms may also originate from the pubis or penile shaft. In both these locations the commonest complaint is *rash* which may be irritant or non-irritant. Contact dermatitis is discussed on page 310.

On the pubis, *irritant rashes* are most often due to parasitic infestations (pubic lice, see p. 291 or scabies, see p. 294) and often involve adjacent areas. In scabies the irritation is typically worse at night or when the patient is warm; in louse infestation the patient may notice nits attached to the hairs or the movement of parasites. Folliculitis, furunculosis and fungal infections may also cause irritation. Rarely, other dermatoses such as lichen simplex (see p. 262) or lichen planus (see p. 264) may present as irritant pubic lesions. Injudicious self-application of ointments containing mercury compound may sometimes cause a severe irritant contact dermatitis (see fig. 491); history of such application should be sought in clinically suggestive cases. Anxiety is another cause of pubic irritation; examination usually shows no abnormality but sometimes patchy damage to pubic hairs due to plucking or excoriation may be evident.

Non-irritant pubic rash is a considerably less common complaint. Conditions such as molluscum contagiosum (see p. 330) may be found or the term may be used in reference to warts or sebaceous cysts.

Irritant and non-irritant rash of the penile shaft is usually due to one of the conditions discussed above. Occasionally a patient will attend complaining of vesicular lesions of the penile shaft; these are usually early lesions of herpes genitalis (see p. 316). The frequent complaint made by adolescents of rash on the penis is usually found to refer to the hair follicles and sebaceous glands on the ventral surface on the shaft which have only recently been noticed by the patient (see fig. 17).

Pigmentary changes affecting the pubis or penis are an occasional reason for attendance. Adolescents may become worried when the physiological increase in genital pigmentation at puberty is observed. Depigmentation may follow trauma or infection (particularly in coloured patients) or may be due to lesions of balanitis xerotica obliterans (see p. 285). In some patients pale macular lesions may be due to tinea versicolor (see p. 258); lesions are usually more marked on the upper part of the body but rarely may be found solely on the genitalia. Haemangiomata may occur on the genitalia and be a source of anxiety (see fig. 21).

Perineal symptoms

Vague symptoms of *subcutaneous aching* or *mild discomfort* originating in the perineal region are fairly common complaints. In the great majority of these cases no organic cause can be found and the symptoms can be attributed to anxiety. Similar symptoms of organic origin are most frequently due to prostatitis (see p. 204), and may be associated with mild obstructive urinary symptoms. Genital herpetic infections are also prone to cause mild perineal discomfort. Rarely, acute prostatitis or acute Cowperitis may cause severe deep perineal pain; perineal symptoms occasionally occur in cystitis or may be due to referred pain from proctitis or anal lesions.

Lumps on the perineum are another occasional complaint. These may be found on examination to be warts or sebaceous cysts (which may be infected) or in very rare cases to abscesses originating in the urethra, prostate or Cowper's gland and pointing on the perineum.

Perineal irritation can be due to parasitic infestation, fungal infection or to any intertriginous dermatosis. The possibility of contact dermatitis should not be forgotten.

Abnormalities of erection

In practice, *inadequate, absent or rapid loss of erection* is the commonest complaint in this category. Anxiety or other psychological causes are by far the most likely cause of this symptom; patients may attend a venereo-logical clinic with a belief that the phenomenon is related to previous sexually transmitted disease.

Pain on erection or with ejaculation may occur in acute urethritis or may be due to trauma sustained during sexual activity. Either of these symptoms may sometimes be the initial complaint in urethritis but in most cases the history and other findings make assessment of the complaint easy.

Curvature of the penis during erection is another symptom which may cause attendance. In the majority of cases the complaint has no pathological significance but a small number of patients are found to have Peyronie's disease (see p. 338).

Priapism (persistent penile erection) is a symptom that occurs occasionally, seldom alone. It is usually due to urethritis or trauma (often resulting from sexual activity) but in rare cases may be due to leukaemia or to a crisis in sickle cell anaemia.

Bleeding

In practice, complaint of frank bleeding is unusual but blood-staining may be noticed with urethral discharge, ejaculation or as stains on clothing. Frank bleeding is nearly always associated with trauma, either accidental or occurring during sexual activity. The possibility of trauma associated with masturbation should not be forgotten. The commonest cause of bleeding is a tear of the fraenum causing haemorrhage (which is occasionally profuse) from the fraenal artery and vein. Traumatic haemorrhage from other genital sites may also occur; history is usually diagnostic.

Scanty blood at the meatus may be the presenting symptom of intra-urethral warts; warts in other areas occasionaly bleed after mild trauma. With severe urethritis and cystitis bleeding may occur, almost always in association with urethral discharge or urinary symptoms.

Blood, often rusty or brown in colour, may be noticed in the ejaculate (*haematospermia*); this is a fairly frequent symptom in prostatitis (see p. 204) and may be the initial presentation. Haematospermia is a rare presentation of hypertension. The blood pressure should always be checked in patients with this symptom.

A small number of patients attend complaining of blood stains on the underwear. The cause may be one of those discussed above or it may be due to the minute bites of pubic lice; in the latter case examination of the under-wear shows multiple pin-points of dried blood.

FEMALE PATIENTS

Introduction

Female patients may attend with symptoms or, more frequently, are totally asymptomatic and attend as a contact of a sexual partner already found to have transmissible disease. The evaluation of genital symptoms and the examination of asymptomatic attenders must always be comprehensive and investigation with appropriate tests of all locations where evidence of disease may be found must be carried out. The findings are seldom so clear-cut as to make diagnosis possible on clinical grounds alone, as the same symptoms and morphological appearance may be observed in conditions of differing aetiology. Even when symptoms are present subjective factors in the individual patient influence the degree of complaint; findings on examination that appear identical to the doctor may be regarded as normal by one woman and abnormal by another. Further sources of confusion are lack of specificity in symptoms, the frequent finding that disease may affect several anatomical sites simultaneously, and the terminology employed by patients. Paradoxically, it is also quite common for symptoms only to be noticed by their absence—after treatment the patient is aware of a change due to symptoms, previously ignored, having disappeared.

In this subsection the significance of some of the more frequent symptoms is discussed in more detail. Inevitably there is considerable overlap. Concise tabulation of the aetiology of symptoms in female patients is difficult because of the mixture of symptomatology, anatomy and pathology. My own attempts at the analysis of symptoms are open to criticism from several aspects but are presented as indications of the aetiology of the more frequent complaints in adult patients. The conditions mentioned are more fully described later. No attempt has been made to include genito-urinary disorders mainly occurring in paediatric or gynaecological practice. Ulcerative conditions have been discussed on page 24, lumps in the groin on page 26 and anogenital symptoms on page 28.

Bleeding

The most usual causes of the complaint of bleeding are trauma or lesions of the cervix. Less commonly, bleeding may be due to gynaecological disorders (which are outside the scope of this book) or lesions of the vulva, urinary tract or vagina. The history is usually diagnostic when the bleeding is due to trauma. Vulval ulceration, urethral caruncle, acute vaginitis and acute urinary tract infections may be the cause of the bleeding and with these conditions other symptoms are usually present. Erosion of the cervix is frequently observed and may cause inter-menstrual or post-coital bleeding; other cervical lesions such as ulceration, endometriosis or polyp may cause similar symptoms but are less common. Contraceptive methods must always be considered in any patient with this complaint—oral contraceptive agents and intra-uterine devices are both fairly frequently associated with menstrual irregularity and inter-menstrual spotting. It should not be forgotten that patients may use instruments or corrosive agents on themselves (often to procure miscarriage) and in these circumstances a truthful history may not be forthcoming.

Dyspareunia

Pain on intercourse experienced at the introitus (*superficial dyspareunia*) is often associated with painful vulvitis, which is most likely to be due to candidiasis, trichomoniasis or herpes. Other vulval lesions, such as inclusion cysts and hymenal tags, may also be a cause of the symptom. Deficiency of the physiological mucous secretions may make intercourse painful—the deficiency may be due to emotional factors, to excessive washing or to over-hasty intromission. When intercourse is attempted the resultant pain is likely to exacerbate the aridity.

Tenderness of the vaginal walls during intercourse may occur in patients with vaginitis. Pelvic pain on intercourse (*deep dyspareunia*) is a complaint of patients with salpingitis and endometritis or it may occur with gynaecological disorders such as ovarian cyst. The 'chandelier' sign, which is positive when pelvic pain is provoked by movement of the cervix during bimanual examination, is a feature of deep dyspareunia which is useful in the examination and evaluation of patients with complaint of pelvic pain.

Intra-uterine contraceptive devices occasionally cause mild deep dyspareunia, and often cause colicky discomfort during menstruation.

Discharge

Discharge is a frequent compliant usually referring to an alteration of genital secretions which is regarded as excessive (a very subjective term) or abnormal by the patient. Discharge may be vulval, vaginal or cervical in origin and the cause may be physiological or pathological. In the history and examination, points to be particularly noted are the amount, colour and consistency of the discharge, the presence of abnormal odour, the presence of associated symptoms and relationship to the menstrual cycle or other specific factors such as sexual intercourse. Personal habits such as vaginal douching and antiseptic usage may cause discharge and relevant enquiry should be made. Evaluation of these points can be of great help in assessment but examination is always necessary to determine aetiology. There are numerous causes of discharge, but in practice the majority of patients are found to have candidiasis, trichomoniasis, gonorrhoea or non-specific genital infection. *Irritant discharge* is particularly associated with candidiasis; discharge with a *'musty'* or *'foul' odour* is most often due to trichomoniasis or a vaginal foreign body; *blood-stained discharge* may be associated with disorders such as cervical erosion or metropathia, and may sometimes occur in acute trichomoniasis.

**Aetiology of increased
discharge in women**

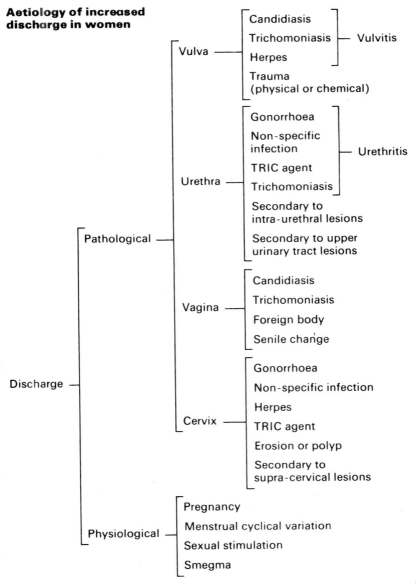

Vulval symptoms

Symptoms related by patients to the vulval region may, on analysis, be found to originate from the pubis, groin, perineum, upper thigh or vulva; a careful history is essential. Frequent symptoms are *irritation, swelling, pain* or *'rash'*. *Irritation* (pruritus vulvae) of the labia majora or pubis is usually due to parasitic infestation; pruritus is almost invariable in symptomatic candidiasis but may be due to any of the other causes of vulvitis. *Vulval swelling,* which is often exacerbated by coitus, is commonly due to candidiasis but may be asociated with trichomoniasis, syphilitic chancres or trauma. *Vulval pain* (and *superficial dyspareunia, q.v.*) may be intermittent or constant and is often related to urination or sexual intercourse. Vulvitis (particularly herpetic vulvitis) is the usual cause but lesions such as Bartholinitis may be responsible. The term *'rash'* may refer either to inflammatory conditions of the skin or mucous membrane or, commonly, is a complaint made by patients who are found to have genital warts. Vulval symptoms may also extend to the perivulval areas such as the groin, perineum and natal cleft in candidiasis and on to the upper thigh in trichomoniasis. Vulval symptoms may also be due to trauma or contact dermatitis, often due to careless use of antiseptics; specific enquiry should be made as the relevant history is often not volunteered. All these symptoms are highly variable in degree and should not be regarded as more than diagnostic clues as none are pathognomonic.

'Lumps' of the vulva and adjacent areas are a frequent cause of complaint. The lumps may be due to physiological structures discovered during self-examination — hair follicles, sebaceous glands, varicosities and hymenal tags are examples of causes of this complaint which are not pathological conditions but may be a source of anxiety. Painless lumps such as vulval or perivulval haemangiomata, epidermal inclusion cysts, Gartner's duct cysts, and Bartholin's cysts may also cause the patient to seek advice. Cystocele, cysts of Skene's glands and urethral lesions such as caruncle or prolapsed mucosa may be found to be the reason for the complaint. Genital warts (see p. 332) are frequently first noticed as lumpy growths on the labia or vulva. Bruises resulting from trauma (sometimes during intercourse) may be noticed as painful lumps, and other painful lesions include pyogenic skin conditions such as furunculosis or abscess.

Aetiology of vulval symptoms

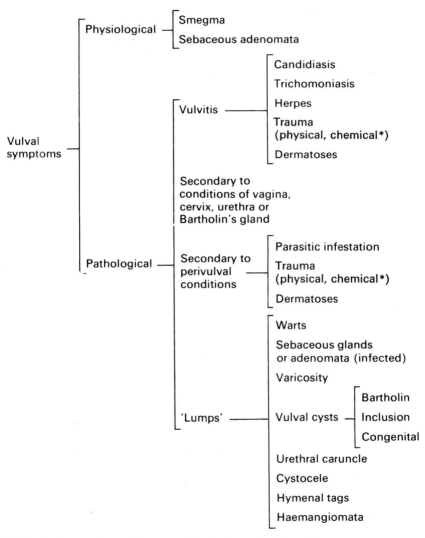

Physical, eg nylon underwear, chemical, eg deodorant spray.

Urinary symptoms

Urinary symptoms in venereological practice are very variable and are never pathognomonic. Complaint may be made of *irritation* or *pain* associated with urination, *upper urinary tract symptoms* and, rarely, retention. In addition, patients often complain of an alteration in *urinary odour* (the term 'strong' is often used) or may notice change in colour which may be due to alteration in specific gravity, blood (haematuria), or may be due to metabolites of drugs which are excreted in the urine; complaint of green urine is usually due to ingestion of drugs containing methylene blue. Symptoms may be due to lesions of the vulva, urethra, bladder or upper urinary tract; a detailed history is most helpful in analysis. *Pain* or *irritation* on urination confined to the meatus or vulva is usually associated with vulval lesions, urethral caruncle or urethritis, whereas the same symptoms associated with frequency, urgency or strangury are more likely to be due to cystitis or upper urinary tract infection or, occasionally, to referred symptoms in patients with cervicitis. *Severe dysuria* and *acute retention* may be due to urethral or vulval ulceration and is most commonly seen in first attacks of herpes. Appraisal of the complaint of *haematuria* must also consider other origins for blood seen at the vulva or in the voided urine, such as menstruation or bleeding lesions of the vagina or supra-vaginal structures.

Skin symptoms

The majority of symptoms originating from vulval cutaneous tissues have already been discussed in the paragraphs referring to discharge and vulval symptomatology but some patients present with symptoms originating on the pubis. The aetiology of this symptom is similar in male and female patients and the reader is referred to page 37. In practice, it is noticeable that females appear much less likely to become anxious about skin symptoms when these are the sole abnormality present; the majority of female patients who mention skin symptoms have other symptoms which are the main reason for attendance. An occasional reason for the complaint of 'rash' of the vulva are the minute, branny, sebaceous glands of the glabrous inter-labial surfaces which may be found (and considered abnormal) by patients during self-examination.

Aetiology of urinary symptoms

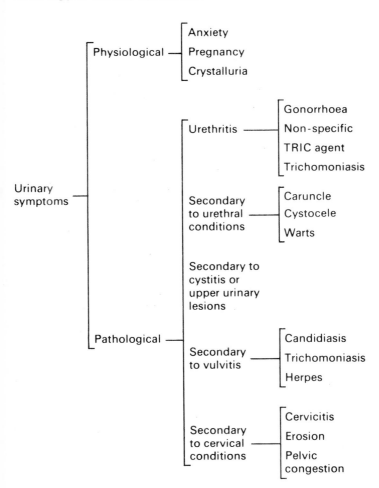

Odour

Complaint of abnormal genital odour is frequently made and may be the only complaint but more commonly is associated with other symptoms such as vaginal discharge. The abnormal odour may be described as *'sour'* or *'acid'* when candidiasis is present; a *'foul'* or *'musty'* odour is typically associated with trichomoniasis. Retained vaginal foreign bodies (often tampons) and necrotic lesions of the vagina or cervix may cause extremely unpleasant 'faecal' odour.

Abdominal pain

Abdominal pain due to lower genito-urinary tract disease is a relatively infrequent complaint in practice but is occasionally the sole reason for presentation. It must always be remembered that patients who are anxious about 'V.D.' may attend a clinic with abdominal pain which is due to a non-venereal cause. *Mild aching* and *discomfort* in the lower abdomen and groins may occur with any genito-urinary pathology but is common in patients with acute vaginitis or herpetic infection. Cystitis (which may be due to infection ascending from the urethra) is another common cause of lower abdominal pain encountered in practice; pain is usually supra-pubic and intermittent and is often exacerbated by urination. Salpingitis may be the presenting feature in patients with gonorrhoea or non-specific genital infection; sometimes pain is only felt on intercourse (*deep dyspareunia, q.v.*) but more commonly the pain, of variable severity, is felt in one or both iliac fossae and is colicky and intermittent, but may be persistent. In patients with endometritis, pain is usually central and may be referred to the rectum or anus or to the groin: deep dyspareunia is usual. Assessment of the complaint of pain requires careful analysis of the history and examination findings.

TECHNIQUES OF EXAMINATION

INTRODUCTION

Many people attending a venereological clinic for the first time are, understandably, extremely apprehensive and it is essential that the rooms in which patients are interviewed and examined have quiet, privacy and good illumination.

Full clinical examination will often be necessary as systemic disease is often the cause of genital symptoms and lesions. Complete genital examination is always necessary and particular attention should be directed to remote sites where lesions that are helpful in the diagnosis of genital disorders may often be found. These remote sites include the eyes and conjunctivae, the oral cavity, the skin and nails and the lymphatic system. Examination must always include taking specimens for serological examination for evidence of syphilis; in many other conditions further laboratory examinations will be necessary.

Genital examination of male patients

The external genitalia, pubis and inguinal regions are inspected for superficial lesions or glandular enlargement.

The shaft of the penis is palpated, and the condition of the testes, epididymides and spermatic cords is assessed by palpation.

The prepuce (if present) is examined and retractibility assessed. In some patients contraction of the preputial margin (phimosis) may make reflection of the prepuce impossible; in rare cases surgical incision of the prepuce (dorsal split) may be necessary to allow adequate examination.

Examination of the glans penis, the coronal sulcus and sub-preputial sac is made with the prepuce reflected: if balanitis or balano-posthitis is present, bacteriological specimens are taken.

The external urinary meatus is examined. If urethral discharge is present it may be visible or expressed by digital massage along the line of the urethra. It is preferable that urine should be retained for several hours before urethral examination — if urination has been recent urethral discharge may not have had sufficient time to accumulate. Suitable bacteriological specimens are taken.

The perineum, anus and perianal regions are inspected. In passive homosexuals proctoscopy and bacteriological examination is necessary. Digital examination of the prostate gland may be required, but this procedure is usually performed later.

At the conclusion of examination the patient should be requested to pass urine for the 'two glass' test (see p. 63) and for testing to exclude abnormal urinary constituents.

39 The abdomen, genitalia, groins and thighs are inspected.
40 The prepuce is retracted.

39

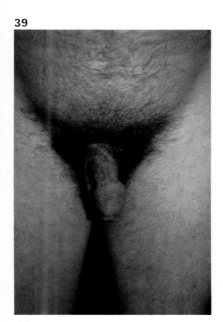

40

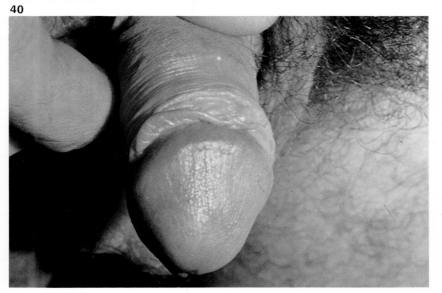

51

Genital examination of female patients

Examination is most easily conducted with the patient in lithotomy position. Good lighting, with an adjustable lamp, is essential.

The external genitalia, pubis and anal region are inspected and the groins palpated. Any enlargement of Bartholin's glands is noted and the presence of urethral or vaginal discharge is also noted. The labia should be carefully separated and the clitoral hood reflected for adequate examination.

A speculum of suitable size and design (*e.g.* Cusco) is gently introduced through the introitus and passed into the vagina to expose the cervix uteri. If lubrication of the instrument is necessary to facilitate introduction this should be done with water or saline as the use of other lubricants may interfere with subsequent bacteriology. The vaginal walls and cervix are inspected and bacteriological specimens taken. Occasionally, removal of foreign bodies is necessary.

After dry-swabbing of the cervix and inspection, endocervical bacteriological specimens are taken, and smears for cervical exfoliative cytology may also be taken. The latter specimen may be taken at initial examination but in many clinics this is not done until any infection present, which may make interpretation of the smear difficult, has been treated.

After withdrawal of the speculum the urethral meatus is inspected and bacteriological specimens taken. If findings indicate the need, bacteriological specimens are taken from lesions of Bartholin's glands or Skene's glands.

Proctoscopy and examination of rectal smears and cultures may be helpful, but this procedure is usually not done as a routine measure. This investigation may be particularly helpful in patients suspected of gonorrhoea or threadworm infestation.

Bimanual pelvic examination is undertaken to assess the uterus and other pelvic organs. Inspection and palpation of the abdomen concludes the examination.

The patient is requested to pass urine, which is then tested for the presence of abnormal constituents or may be sent for laboratory examination.

41 Inspection of the urethra.
42 Inspection of the vulva. The patient has herpes.
43 Palpation of the groin for enlarged glands.
44 Inspection of the cervix and vaginal vaults.

41

42

43

44

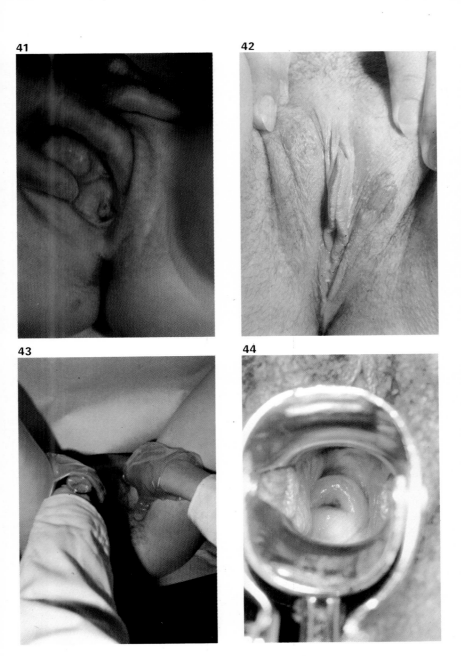

Urethral and sub-preputial examination of male patients

Apparatus required Microscope slides and cover slips. Platinum loop or other suitable swab, and culture swabs. Culture and/or transport media. Saline. Cleansing swabs.

Technique The prepuce (if present) is reflected. In cases of balanitis or balano-posthitis the specimens are taken from the preputial sac. In suspected urethritis the glans penis is cleaned and the specimens taken from the urethral orifice with the platinum loop. For stained smears, the specimen is spread thinly on a glass slide and allowed to dry. For smears to examine for *Trichomonas vaginalis* the specimen taken is mixed with a drop of saline on the slide and the drop is then covered with a cover slip.

Cultures are taken with suitable swabs. The specimens may either be inoculated directly onto or into a suitable medium or the specimens may be placed in transport medium for transmission to the laboratory.

Ulcerated lesions usually require examination by dark-ground microscopy; the technique is described on page 58. Occasionally other diagnostic methods are used.

45 Taking an urethral smear with platinum loop.

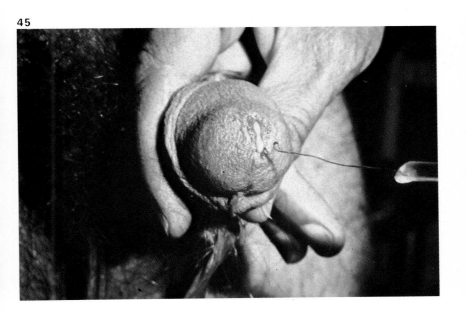

Urethral, vaginal, cervical and rectal examination of female patients

Apparatus required As for urethral examination in male patients (see p. 54). Vaginal specula. Sponge-holding forceps. Proctoscope and lubricant.

Technique The vaginal speculum is passed to expose the vaginal vault and the cervix ; the vaginal walls are inspected. With a platinum loop, material is taken from the pool of vaginal secretion in the posterior fornix — this material is used to make preparations for staining and for examination for *Trichomonas vaginalis.* The external surface of the cervix is cleansed with a dry swab and specimens for staining are taken from within the cervical canal.

Specimens from the urethra are taken in a similar manner ; usually a specimen for staining is the only requirement but drop preparations for *T. vaginalis* may be necessary. If specimens are required from Bartholin's glands or Skene's glands these are prepared in the same way.

Proctoscopy is performed to inspect the anal canal and rectum and for specimens to be taken.

Cultures are taken from the same sites and are either plated directly onto or into suitable media or placed in a transport medium.

Ulcerated lesions usually require examination by dark-ground microscopy or other methods (see p. 58).

46 Introduction of Cusco speculum.

46

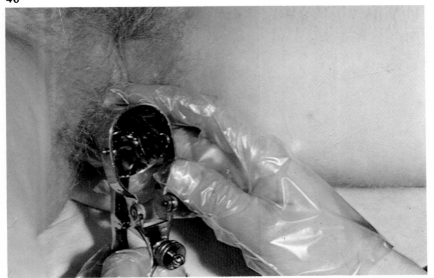

Ulcerated lesions

The aetiology of ulcerated genital lesions is protean but early syphilis is the most important disease to exclude. In this disease the microscopical examination of specimens with *dark-ground* (or *darkfield*) illumination is essential. Other diagnostic methods (*e.g.* other bacteriology, biopsy etc.) may be employed but these are not detailed here. The examination may have to be repeated on several occasions or on several consecutive days before a firm diagnosis can be made: no treponemicidal therapy should be used until examinations are completed.

Apparatus Glass slides and cover slips (thin slides are most suitable). Scarifiers: a cotton swab is often suitable but a metal instrument may be used. Capillary tubes.

Technique The lesion is cleaned with normal saline to remove surface debris. The surface is scarified and then the base of the ulcer gently squeezed so that serum exudes onto the surface. This serum is collected on the edge of a cover slip, which is then placed in the centre of a slide. The preparation is compressed to produce a thin film which is immediately examined microscopically. If microscopic facilities are not immediately available, serum may be collected in a capillary tube. The ends of the capillary tube are sealed and the tube sent to the laboratory for examination; in such preparations *Treponema pallidum* can survive several days.

Gland puncture

This technique may be used to obtain material for dark-ground microscopical examination in cases when such material is unobtainable from the ulcerated lesion but enlarged regional lymph nodes are present. A similar technique is sometimes employed to collect material from the edge of healing ulcers.

Apparatus Syringe and intramuscular needle. Thin glass slides and cover slips. Sterile saline. Skin disinfectant.

Technique The skin is disinfected. A minute quantity of saline is aspirated into the needle. The needle is introduced into the gland and the

47 Darkfield microscopy: scarifying lesion with swab.
48 Darkfield microscopy: scarifying lesion with metal scarifier.
49 Darkfield microscopy: collecting serum in capillary tube.
50 Darkfield microscopy: collecting serum on cover-slip.

47

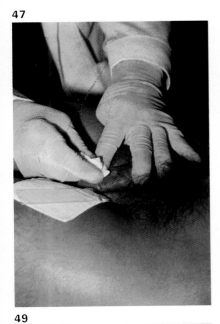

48

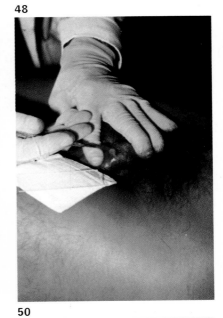

49

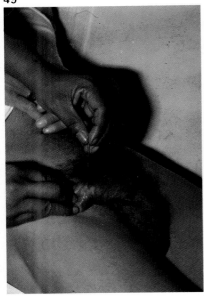

50

saline expelled. The gland is gently massaged over the tip of the needle and the syringe is aspirated. After withdrawal from the gland the material in the needle is expelled onto a slide and covered with a cover slip. The preparation is compressed and examined microscopically.

Proctoscopy

Proctoscopy is essential in the examination of male passive homosexual patients and is often used in the diagnosis of gonorrhoea in female patients. In addition, proctoscopy may be helpful in the evaluation of patients suspected of having syphilis, lymphogranuloma venereum or genital warts, and occasionally in other conditions.

Apparatus Proctoscope (an instrument with built-in illumination is preferable). Lubricant. Platinum loop and culture swab. Culture and/or transport media. Glass slides and cover slips.

Technique Female patients are examined in lithotomy position. Male patients may be examined in either a kneeling position or the lateral position. The lubricated proctoscope (with obturator *in situ*) is gently pressed against the anus and inserted into the rectum; it is often helpful if the patient is asked to push backwards against the instrument during introduction. The obturator is removed and the rectum inspected: bacteriological or other specimens may be taken. The instrument is slowly withdrawn while the distal part of the rectum and anal canal is observed.

51 Proctoscopy.

51

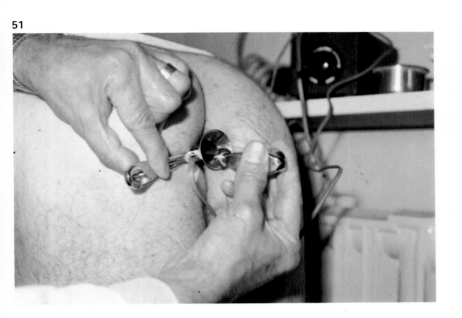

Urine tests

All patients should have urine specimens tested for the presence of abnormal constituents such as protein, glucose and blood. Microscopical examination of urine may be necessary to determine the nature of crystal deposits which may occasionally cause symptoms of urethritis. In rare cases ova of *Schistosoma haematobium* may be found. In cases of suspected upper urinary tract infection, mid-stream or clean specimens of urine should be sent to the laboratory for microscopy and bacterial culture.

Male patients

The *'two glass'* test is a useful clinical indication in the assessment of the site of infection and in examination after treatment of urethritis.

The patient is given two urine glasses and requested to pass 50—60 ml. into the first glass and the remainder into the second glass. Debris from any anterior urethral inflammation present will be voided with the first specimen and the second glass will contain urine representative of the contents of the bladder. The test is not completely reliable and the findings may be unrepresentative in very severe cases of anterior urethritis and in cases when the urine has only been retained for a short time before testing.

TECHNIQUES OF EXAMINATION

Two glass test

Hazy in both = phosphaturia or
infection proximal to anterior urethra

Acidify with acetic acid

Hazy or shreds in first glass,
clear in second
= anterior urethritis

or

Clear in both glasses
= phosphaturia

or

Unchanged = infection proximal
to anterior urethra
(often + anterior urethritis)

or

Few shreds in first glass,
haze and specks in second glass
= suggestive of posterior urethritis,
prostatitis

The prostate gland and prostatic massage

Examination of the prostate may reveal chronic prostatitis or the now rare acute prostatitis or prostatic abscess. Cowperitis and vesiculitis may also be detected by rectal examination, but these are also rare conditions.

Examination of the prostatic (or prostato-vesicular) secretion expressed by digital massage of the gland was formerly a common procedure in tests of cure after treatment of urethritis. At present the technique is still occasionally used for this purpose but is more frequently employed in the investigation of persistent urethritis or other genital symptoms and in conditions such as uveitis and Reiter's syndrome, or in treatment of chronic prostatitis.

The significance of abnormal findings in prostatic fluid remains a matter of controversy (see p. 204).

Apparatus Finger stall and lubricant. Glass slides and cover slips. Culture and/or transport media (rarely required).

Technique of prostatic massage The patient should be either in the knee-elbow position or bending over a low chair. A finger, in a lubricated finger stall is introduced into the rectum and the prostate and seminal vesicles palpated. Prostatic fluid is expressed by digital massage (see diagram) and collected at the urethral orifice for microscopic or bacteriological examination. It may be necessary to 'milk' the urethra to express the fluid. Massage should not be done in the presence of acute prostatic infection because of the risk of bacteraèmia.

Diagrammatic anatomy

Prostate massage

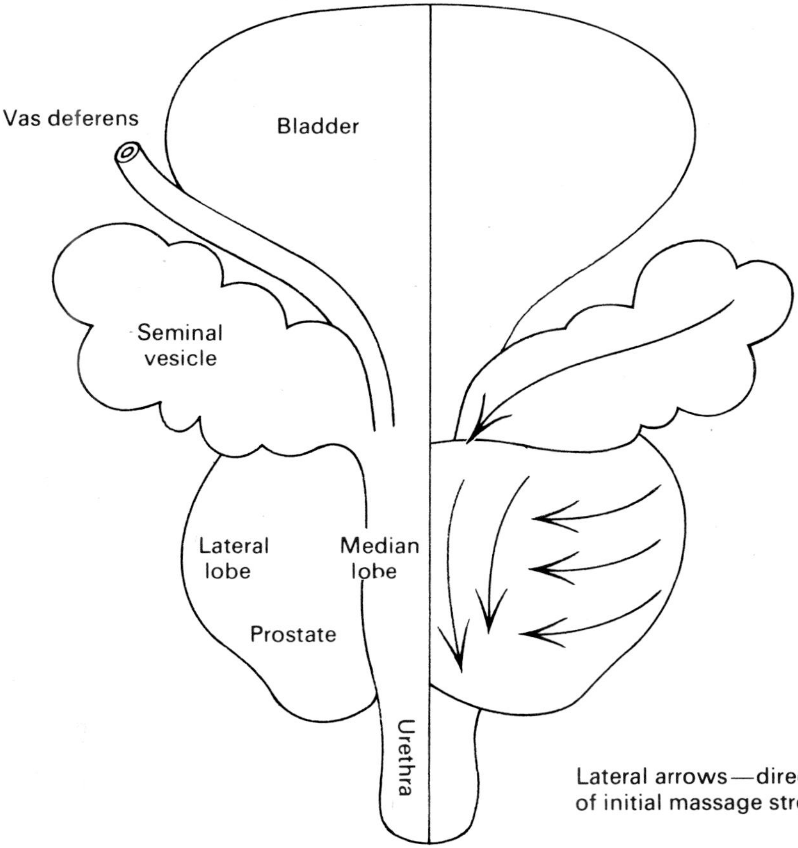

Vas deferens

Bladder

Seminal
vesicle

Lateral
lobe

Median
lobe

Prostate

Urethra

Lateral arrows—direction
of initial massage strokes

Medial arrows—direction
of final massage strokes

Anterior urethroscopy

Direct visual examination of the anterior urethra is most often used when urethral stricture or urethral infiltration is suspected. The technique may also be used in the investigation of cases of persistent urethritis, or when Littritis or another intra-urethral lesion is thought to be present. Occasionally intra-urethral surgical techniques may be performed with the urethroscope. In the past, anterior urethroscopy to exclude stricture was a part of the tests of cure after treatment of urethritis but this complication is now so rare that routine urethroscopy is seldom practised. The instrument may also be used for vaginal examination in infants.

Apparatus Anterior urethroscope (the instrument consists of a cannula with obturator to which may be attached a lighting system, a telescope and air distension system). Urethral anaesthetic. Penile clamp. Disinfectant.

Technique Strict sterile precautions must be observed and, with rare exceptions, the examination is not carried out while acute urethritis is present. After cleansing and sterilisation of the glans penis local anaesthetic is introduced and massaged down the urethra. The penile clamp is applied until the anaesthetic has taken effect. The lubricated cannula of the urethroscope (with obturator in place) is gently passed down the urethra as far as it will comfortably go. The obturator is withdrawn and the light and telescope fittings are attached. After focusing the telescope the urethra is distended by air pressure and the epithelium observed; normal epithelium has a glistening and pink appearance. The urethroscope is slowly withdrawn while the walls are kept under observation. Oedema or strictures may be found; when Littritis is present the intra-urethral duct orifices are infected and swollen, and show up as dark dots.

52 Urethroscopy.

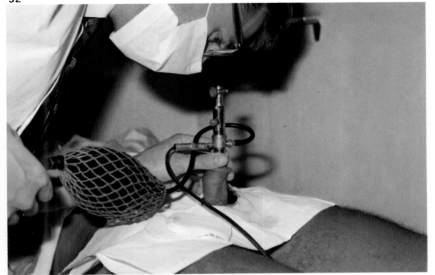

Passage of urethral sounds

Urethral sounds are most often used for the treatment of stricture of the urinary meatus or urethra by dilatation but may also be used in the diagnosis of urethral stricture. Palpation of the urethra over a straight sound may reveal small firm nodules in cases of Littritis. The routine passage of urethral sounds as part of the tests of cure after urethritis is now seldom practised as the complication of stricture is now so rare.

Metal urethral sounds may be either straight or curved and are graduated in size. The straight sounds are for anterior urethral use only, whereas the curved sounds may be passed through to the bladder. Urethral bougies (formerly gum elastic, now usually plastic) may be employed in a similar manner.

Apparatus Sets of graduated straight and curved metal sounds. Sets of bougies. Urethral anaesthetic. Penile clamp. Disinfectant.

Technique Strict sterile precautions must be observed and the technique is contra-indicated when acute infection is present. The glans penis is disinfected and urethral anaesthetic introduced. A sound of suitable size is selected and lubricated. For diagnostic purposes English gauge 9 (French gauge 18 or 14/18 for curved sounds) is usually suitable. For treatment the initial size selected will depend on the lesion present. The penis is held vertical and the sound allowed to slide into the urethra without the application of more than the gentlest pressure. With curved sounds it is necessary to rotate the tip of the instrument to pass it through the prostatic urethra into the bladder.

Urethral stricture will obstruct passage of the sound. Dilatation of strictures is accomplished by passing sounds of increasing size until normal urethral dimensions are reached. It is advisable for a patient with a stricture to attend periodically for assessment of the stricture.

Collection of blood specimens

It is essential that serological tests for syphilis (S.T.S.) be performed on all patients attending venereological clinics. These tests are most important in the diagnosis and management of the disease. In other conditions appropriate blood tests may be helpful, particularly complement-fixation tests in lymphogranuloma venereum and genital herpes.

Apparatus Sterile dry syringes and needles. Collecting tubes. Tourniquet. Skin disinfectant. Stylets for heel stabs in infants.

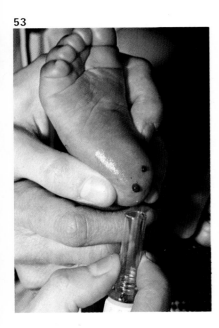

Technique In adults it is usually easy to collect blood by venepuncture in the ante-cubital fossa. A tourniquet is placed round the limb proximal to the vein selected. The skin is disinfected. The vein is steadied and the needle inserted. Blood is aspirated into the syringe and the tourniquet is released before the needle is withdrawn. Digital pressure will control any bleeding from the puncture. The blood in the syringe is expelled into the collecting tube after the needle has been removed : if the blood is expelled through the needle haemolysis may occur and invalidate subsequent examination.

In infants it is easiest to collect specimens by heel stab. The stylet is employed and the specimen collected in a capillary or other collecting tube. Other sites for venepuncture include scalp veins and the central sinus.

53 Heel stab in infant.

Other techniques

Many other methods of investigation may occasionally be used in venereological clinics. These investigations may be indicated by the history given by the patient or by clinical examination.

Potassium hydroxide examination (KOH) Fungal elements such as mycelium and spores may be demonstrated by taking scrapings from suspect lesions; the scrapings are emulsified in a drop of 20 per cent KOH on a slide. The preparation is covered and examined microscopically after autolysis of skin scales has occurred.

Lumbar puncture Indicated in cases of neurosyphilis or to exclude asymptomatic neurosyphilis.

Radiology Indicated in syphilis when skeletal lesions or aortitis are suspected. Indicated in conditions such as Reiter's syndrome or other arthritides.

Skin tests Indicated in suspected contact dermatitis and some other dermatoses, and for diagnostic purposes in lymphogranuloma venereum (Frei test see fig. 259).

Slit-lamp microscopy Indicated in suspected congenital syphilis to exclude previous interstitial keratitis, and in the management of uveitis which may occur in conditions such as Reiter's syndrome.

The investigative methods referred to above are several of the most frequently used; there are, of course, many others that may be used from time to time.

THE TREPONEMAL DISEASES

The treponemal diseases (the *treponematoses*) are a group of conditions which are caused by organisms at present morphologically and serologically indistinguishable but which have different clinical patterns. It seems probable that the clinical patterns observed today result from the adaptation, over the ages, of a common ancestral organism to environmental changes. *Venereal* (sexual) and *non-venereal* transmission may occur. The non-venereal treponematoses are usually designated benign as late complications and congenital transmission are either unimportant, rare or unknown.

Classification
The treponematoses may be classified by the usual mode of transmission:

Venereal transmission *Syphilis.*
Non-venereal transmission *Endemic syphilis, yaws, pinta.*

Many other non-venereal treponematoses have been nosologically differentiated, usually by local dialect names; these conditions probably represent varieties of endemic syphilis (see p. 148).

Distribution

Syphilis World wide.
Endemic syphilis Formerly widespread, now seen almost entirely in areas of low socio-economic status.
Yaws Humid tropical areas
Pinta Central America and north-western areas of South America.

Evaluation of positive serological tests for syphilis (S.T.S.)
In venereological practice one of the common problems that occurs is the
evaluation of the patient with normal findings on clinical examination but
who has positive serological tests for syphilis and originates from an area
where treponematosis is endemic. It is impossible to determine with certainty
whether positive results are residual from treponemal disease acquired in
childhood or are from venereally acquired syphilis. In practice, in the absence
of verifiable information, it is probably best to investigate and treat the
patient as a case of latent syphilis. Recent work with the Fluorescent
Treponemal Antibody (F.T.A.) test employing specific immunoglobulin
fractions suggests that IgM can be found before the clinical lesion appears:
IgG is found shortly afterwards, and persists for long periods. There is some
evidence to suggest that when IgM is found the disease is active. If this
work is confirmed, the significance of the finding of positive S.T.S. will be
easier to evaluate.

54 Electron photomicrograph of *Treponema pallidum*, × 34,000.
55 *Treponema pallidum:* stained foetal liver.
56 *Treponema pallidum:* fluorescent staining.

54

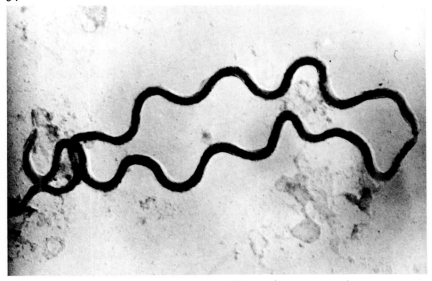

55

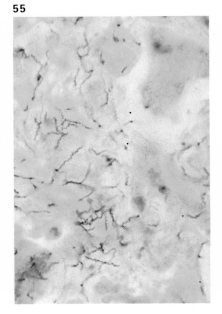

56

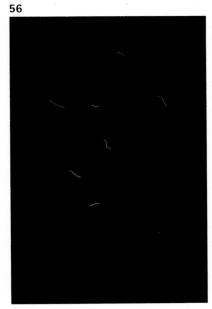

SYPHILIS

Syphilis is a contagious disease caused by *Treponema pallidum*. It is a systemic disease which, when untreated, has overt and covert phases. The diagram below outlines the progress of the untreated infection.

The disease is usually acquired by sexual contact but untreated pregnant women may pass infection through the placenta to the foetus. It is occasionally acquired through non-sexual means (*e.g.* in medical or laboratory workers in contact with infectious patients or the organism, through transfusion of blood from an infected donor to a susceptible recipient and, very rarely, through inanimate objects). Spirochaetaemia and the distribution of the organism throughout the body occur before clinical lesions appear.

The natural history of syphilitic infection

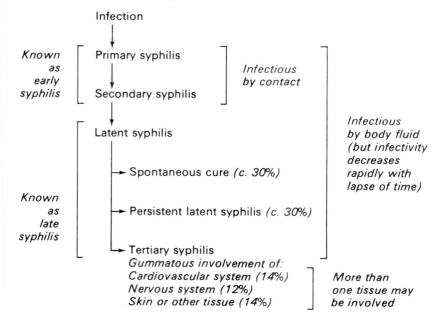

The organism

Treponema pallidum was identified as the cause of syphilis by Schaudinn and Hoffman in 1905. Morphologically the organism is a slender spiral with regular coils and tapering ends. The number of coils is usually between 10 and 15 and the length about 8 μ but organisms may be shorter or longer with less or more coils. The diameter is about 0.25μ. Electron microscopy shows an axial bundle of fine fibrils surrounded by a narrow capsule. Division is usually by transverse fission; it occurs at intervals of 30–36 hours. It seems probable that cyst forms can occur (see figs. 54–56).

Diagnosis of syphilis

The diagnostic methods employed depend on the stage of the disease at the time of presentation; these are discussed in the relevant sub-sections.

Recognition of Treponema pallidum

The only practical method (for clinical use) of demonstrating *T. pallidum* is by dark-ground microscopy. The organism is so slender that when stained by histological methods rapid enough for routine clinic usage it is not differentiated. Microscopy using transmitted light may be used when preparations are stained by silver impregnation. This method is suitable for biopsy specimens but is impractical for routine clinic use. Culture on artificial media is unsuccessful. Experimentally the organism will live and reproduce in suitable animal inoculations but culture is not a clinical practicality. Techniques employing fluorescent staining methods are being developed for clinical use.

Dark-ground microscopy

The technique of preparing a specimen for dark-ground (darkfield) microscopy is described on page 58.

The organism is recognised as a thin silver spiral which is motile, actively angulating and often undulating. The organism may rotate, and the length may vary as the coils compress and expand like a spring. Differentiation has to be made from other spiral organisms (*e.g. Treponema refringens*). The morphological appearances and the patterns of movements of other organisms are different and, with experience, the distinction is usually simple.

Primary syphilis

The lesion is traditionally known as the *primary chancre*. Primary syphilis is the clinical stage of the disease when the infection first becomes manifest as a lesion of skin or mucous membrane. The majority of lesions (95 per cent) occur on or adjacent to the external genitalia. Lesions may be concealed (*e.g.* intra-urethral, rectal, cervical and anal lesions) and the patient is often unaware that infection is present. Other areas of sexual contact (*e.g.* mouth, nipple) may be the location of extra-genital lesions, and other extra-genital lesions may be the result of non-sexual infection in medical and other workers in contact with the disease or organism.

Incubation period
The time between infection and the appearance of the primary lesion is usually between 21 and 35 days, but may occur any time between 10 and 90 days after infection.

Clinical presentation
The primary lesion (*chancre*) begins as a small dusky red macule which soon develops into a papule. The surface of the papule erodes to form an ulcer which is typically round and painless, with a clean surface. The base of the lesion is indurated and feels firm or hard on palpation. Considerable oedema of adjacent tissues is often present. Untreated, the ulcer pursues an indolent course, slowly enlarging to about 2 cm diameter and then slowly healing, usually without residual scarring, after four to eight weeks. In about 50 per cent of cases a single chancre is present; in the remainder, multiple but similar lesions, often 'kissing', are found; confluence may occur. Regional lymph gland enlargement (*syphilitic bubo*) begins one to two weeks after the appearance of the primary chancre. On palpation affected glands are felt to be firm, discrete and slightly to moderately enlarged. Gland involvement may be painless but a proportion of affected patients complain of aching and/or tenderness of the involved glands. In clinical practice about 50 per cent of patients found to have primary syphilis have a syphilitic bubo.

Following treatment the primary chancre heals very quickly: it is uninfectious within 24 hours of starting effective treponemicidal therapy. When a bubo is present when treatment is begun, resolution of the glands may take several months.

57 Typical primary chancre of coronal sulcus.
58 Primary chancre of glans penis.

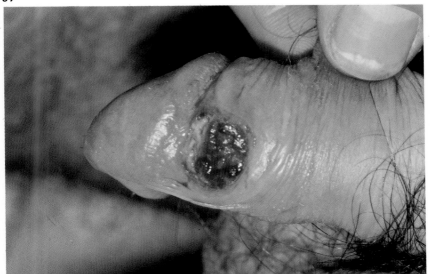

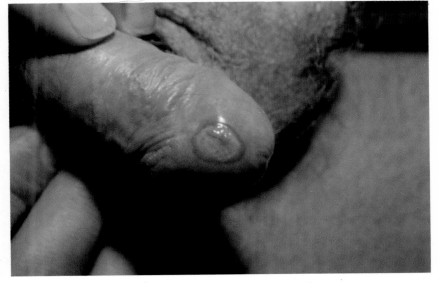

Diagnosis

Diagnosis is made by demonstration of *Treponema pallidum* (see p. 75) by dark-ground microscopy in material taken from the suspect lesion. Clinical diagnosis is inadequate as many primary chancres have an *atypical* appearance and other conditions may mimic primary syphilis.

Serological examination is also essential. About 60 per cent of patients with primary syphilis are found to have positive reactions to reagin tests: in patients who are found to have negative serological tests the infection has been present for too short a time for antibody to reach detectable levels.

59 Extensive ulceration in primary syphilis. Note absence of secondary infection.
60 'Kissing' chancres of coronal sulcus.
61 Multiple primary chancres.
62 Giant primary chancre: 'syphiloma'.
63 Superficial chancre of corona.

59

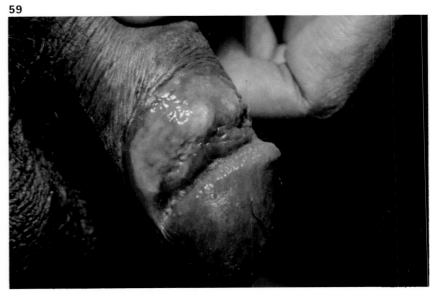

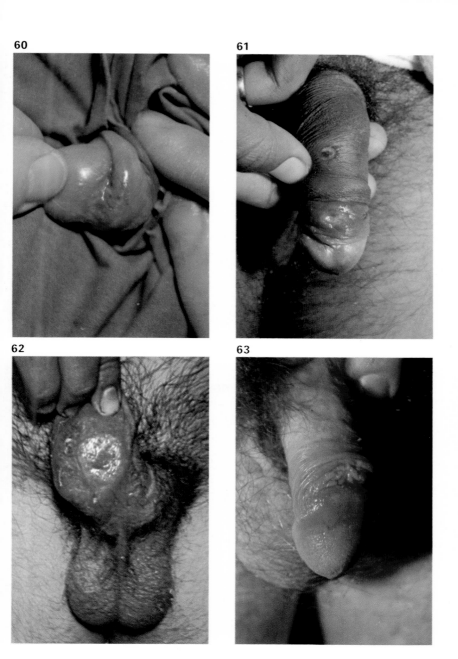

64 Extensive 'kissing' chancres of fraenum.
65 Meatal chancre.
66 Glans chancre: the 'cigarette burn' chancre.
67 Chancre of fissured, phimotic prepuce.
68 Chancre of prepuce. Note extensive oedema.
69 Dorsal view of extensive superficial chancres. The lesions were not indurated (ventral view shown in fig. 70).

64

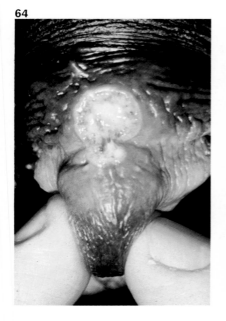

65

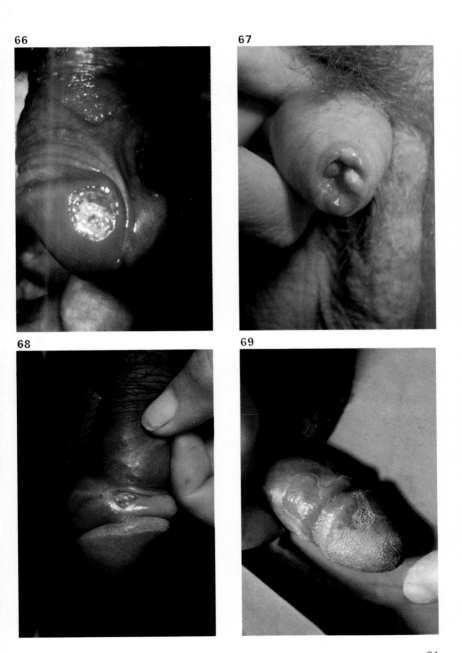

70 Ventral view of extensive superficial chancres. The lesions were not indurated (see also fig. 69).
71 Shaft chancre and bubo of groin.
72 Meatal chancre and syphilitic bubo.
73 Multiple small primary chancres.
74 Minute linear excoriation on shaft. Darkfield positive; very early chancre.
75 Multiple papules. Primary syphilis.

70

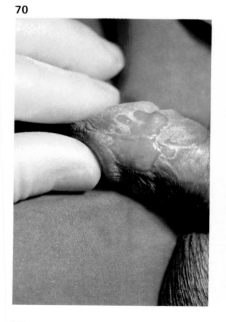

71

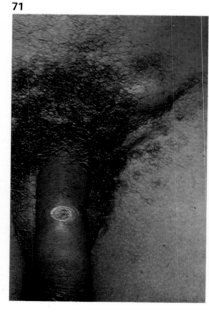

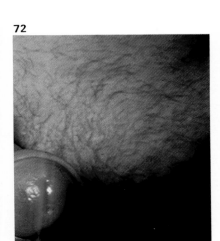

72

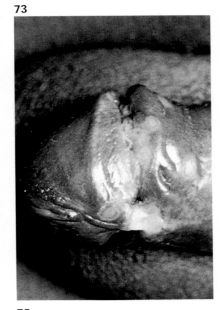

73

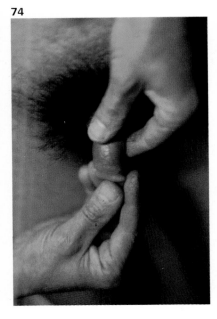

74

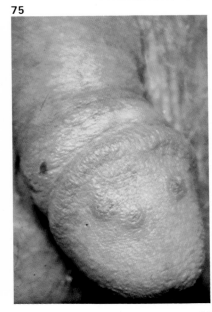

75

76 Penile scarring following primary syphilis: infrequently seen.
77 Multiple small chancres associated with phimosis and secondary infection.
78 Healing 'french letter' or 'condom' chancre at penile root.
79 Another 'condom' chancre.
80 Primary chancre of labium minus. Note also vesicular and erosive herpetic lesions.
81 Superficial primary chancre of labium minus.

76

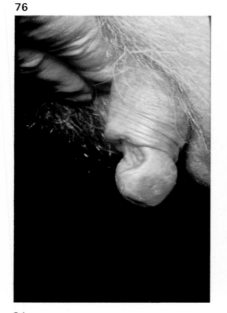

77

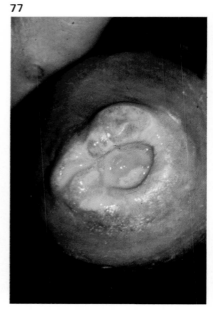

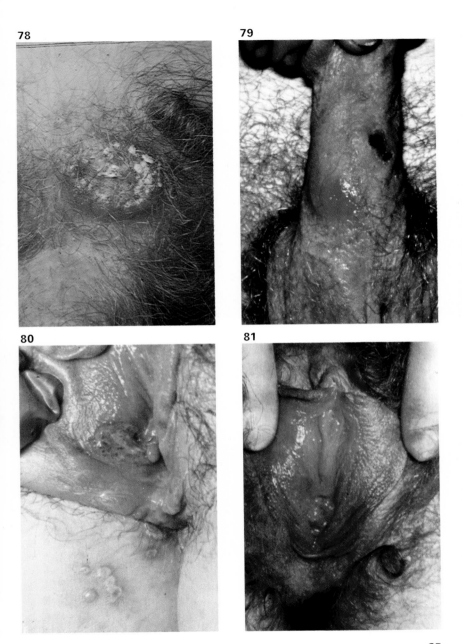

82 Typical chancre of labium majus.
83 'Kissing' chancres of labia majora.
84 Superficial irregular ulceration and oedema of labium majus; darkfield positive. (Cf. fig. 92).
85 Small superficial chancres of introitus.
86 Massive labial oedema overlying vulval primary chancre.
87 Early chancre of cervix. Note similarity to cervical erosions.

82

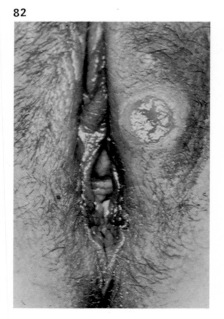

83

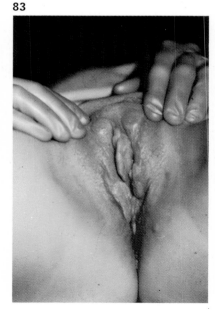

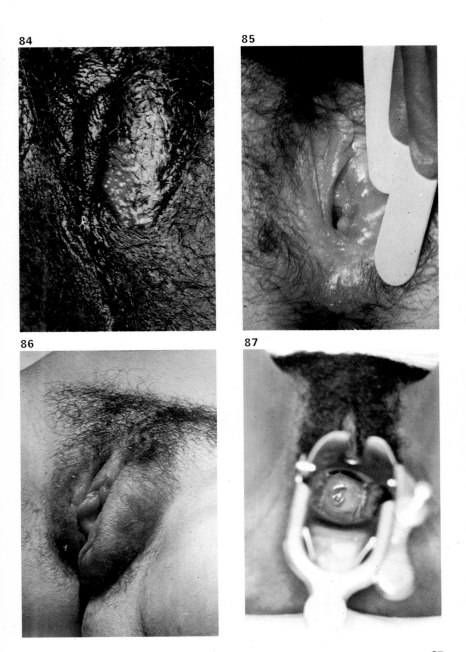

88 Extensive chancre of cervix (note similarity to herpes, figs. 517, 518).
89 Primary syphilis: ulceration on pre-existing anal warts.
90 Primary chancre of buttock.
91 Primary chancre of anus.
92 Primary chancre of anus. Note superficiality of lesions (Cf. fig. 84).
93 Primary chancre of anus.

88

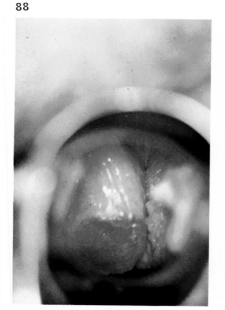

89

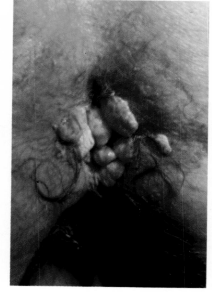

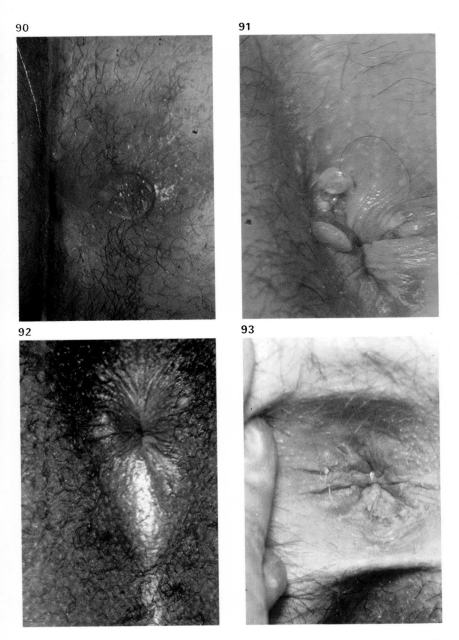

90

91

92

93

89

94 Primary chancre of anus. Note resemblance to anal fissure.
95 Primary chancre of eyelid.
96 Primary chancre of lip.
97 Primary chancre of chest wall.
98 Primary chancre of finger.

94

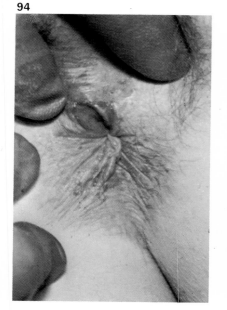

95

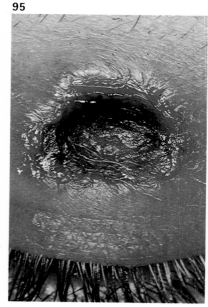

96

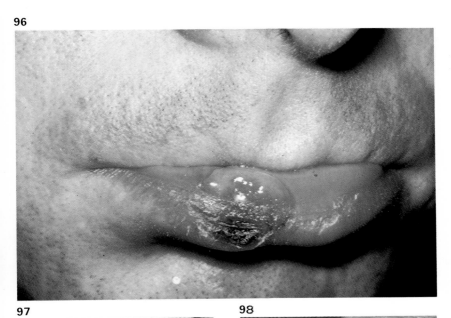

97

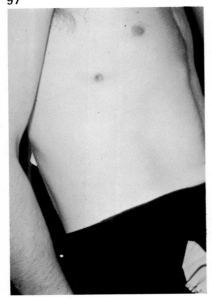

98

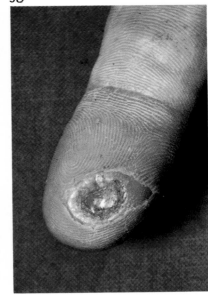

99 Primary chancre of breast. Transmitted from an oral mucous patch.
100 Primary chancre of angle of mouth.

99

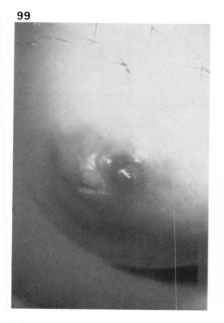

100

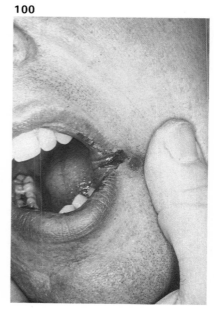

Secondary syphilis

In secondary syphilis the spirochaetaemia which has resulted in widespread dissemination of the organism throughout the body becomes manifest. Lesions may be found on the skin from the scalp to the soles of the feet; lesions may be found on the mucous membranes; there is often generalised lymphadenopathy and mild constitutional symptoms are common. Rarely, visceral involvement may occur. The diagram on page 95 indicates the features of secondary syphilis. In about 30 per cent of patients seen the primary chancre is found to be present; other patients may give a history suggestive of the primary stage but often such history is lacking.

Time of appearance
Signs of secondary syphilis usually appear six to eight weeks after infection. Signs may appear as early as four weeks or, in exceptional cases may not appear for two years.

Clinical presentation
With such a wide range of symptoms and signs the permutations are almost endless. In practice, about 70 per cent of patients with secondary syphilis first attend with a *skin rash*. Other relatively common presentations are *ulcerated lesions of the mucous membranes* or *lymph gland enlargement*. Rarer presentations include falling hair, persistent hoarseness, bone pain, hepatitis and deafness. A considerable proportion (about 30 per cent) of patients who are found to have secondary syphilis attend at the instigation of a sexual contact. Such patients may have been aware of abnormal symptoms or signs but unaware of their significance. The generally mild constitutional symptoms are seldom a sole cause of attendance and only are mentioned when the history is taken. Many of the lesions of secondary syphilis are virtually asymptomatic and are only found in the course of comprehensive and careful examination.

Following treatment, lesions usually heal quickly. Skin eruptions, if they have not been secondarily infected, disappear without residual scarring in the majority of cases but sometimes depigmented areas may be left. Such depigmented lesions are most often seen on the neck. Lymphadenopathy often takes several months to resolve.

Diagnosis
The most important diagnostic method is demonstration of *Treponema pallidum*. Specimens for dark-ground microscopy (see p. 74) may be taken from lesions of the mucous membranes, from condylomata lata, by gland puncture (see p. 58) and occasionally from skin lesions. Serological examination is essential: practically every case of secondary syphilis shows

strongly positive reactions.

It is inadequate to make the diagnosis of secondary syphilis on clinical grounds alone—the range of differential diagnosis is extremely wide. Erroneous diagnosis may result in the unnecessary and stressful attendance of sexual contacts.

Skin eruptions

Skin eruptions in secondary syphilis are characteristically *pleomorphic, symmetrical* and *generalised.* It is rare for the rash to be painful, irritant or vesicular. The lesions are commonly hyperpigmented, the colour varying from pink in the early stages to a dull coppery-red in later stages. Hypo-pigmented lesions (leucoderma) are occasionally seen.

Skin rashes usually first appear as *roseolar* or *macular* eruptions; later *papular* lesions are found. Intermediate (*maculo-papular*) and mixed rashes are common. Superficial scaling over papular lesions is often seen; the lesion is known as the *papulo-squamous* syphilide. Particular patterns of eruption have been given descriptive topographical names in the past *e.g.* annular syphilide, rupial syphilide, corymbose syphilide: these terms are now rarely used. The extremely toxic *pustular* syphilide, characterised by central necrosis of papular lesions, is now very rare in developed countries.

Eruptions in secondary syphilis may be very faint and with few lesions present: recognition can be extremely difficult. In coloured patients hypertrophic lesions are more common, but often the rash is almost invisible. When the eruption is present on the scalp, patchy ('moth-eaten') alopecia may ensue. The rash may sometimes be seen most easily when the patient has been undressed for a few minutes, i.e. when the skin is cool.

Macular and roseolar eruptions

Macular lesions are found mainly on the shoulders, chest, back, abdomen, buttocks, and flexor surfaces of the limbs. The macules are round or oval and 5 to 10 mm in diameter, although sometimes larger lesions are seen which may be due to coalescence of adjacent macules. The intensity of colour of the rash is very variable; it may be necessary for the patient to be undressed for several minutes before the rash can be appreciated. In one recent case the rash was only visible after alcohol had been imbibed.

The features of secondary syphilis

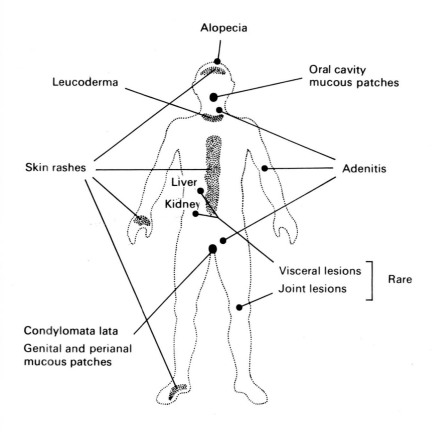

Alopecia

Oral cavity
mucous patches

Leucoderma

Skin rashes

Adenitis

Liver

Kidney

Visceral lesions
Joint lesions

Rare

Condylomata lata
Genital and perianal
mucous patches

Maculo-papular and papulo-squamous eruptions

Papular syphilides are the most frequent and characteristic secondary syphilitic eruption.

The later lesions of cutaneous secondary syphilis show a continuous gradient from the predominantly macular lesion to the predominantly squamous lesion. Initially macular lesions develop central induration to form a palpable nodule; superficial scaling over the papule may occur later. Papules may arise at the sites of macular lesions or may appear on previously uninvolved areas of the skin. Papular eruptions are frequently seen on the face (especially near the hair line: the 'corona veneris') and on the palms of the hands and soles of the feet. The lesions on the palms and soles are often papulo-squamous in appearance. Papulo-squamous lesions may occasionally be seen in other areas, usually those that are subject to friction. Tiny papular lesions occur at the mouths of hair follicles on the scalp, eyebrows, beard area and hairy areas of the trunk. This lesion is known as the follicular syphilide; often the atypical and misleading symptom of pruritis will be present. Papular lesions at the corner of the mouth and the naso-labial junction are often hypertrophied and 'split'. The size of papular lesions may vary from 1–15 mm in diameter; most commonly the size is between 5–10 mm.

Pustular eruptions

Secondary syphilitic eruptions which are predominantly pustular are now rare in developed countries. This type of rash is seen most frequently in Negros and is often associated with debilitation and poor socio-economic conditions. Occasional pustular lesions are quite often seen in widespread maculo-papular eruptions.

Pustular lesions begin as large papules which undergo central necrosis. Severe tissue destruction and toxaemia occur and considerable scarring is likely. On the face, pustular syphilides are often extensively crusted; this lesion is known as the rupial syphilide.

Malignant syphilis, now historical, was a particularly severe form of ulcerated pustular syphilide.

101 Secondary syphilis: macular rash (and one papular lesion).
102 Secondary syphilis: maculo-papular syphilide.
103 Secondary syphilis: psoriasiform syphilide.
104 Secondary syphilis: papulo-squamous syphilide.

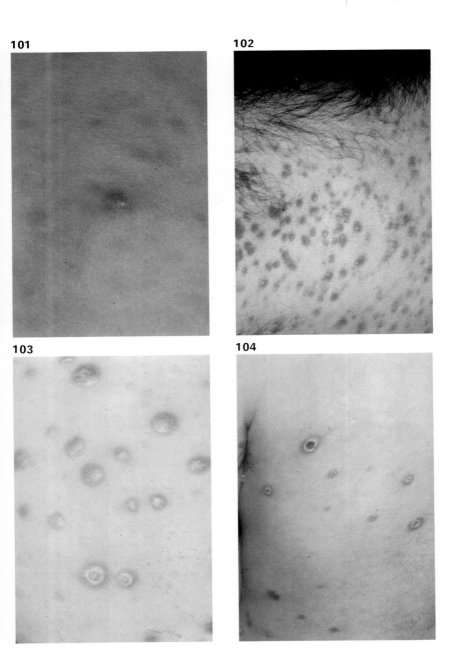

105 Secondary syphilis: purpuric syphilide.
106 Secondary syphilis: plantar syphilide. This rash had been treated for three months with steroids before the true diagnosis was established. Note resemblance to fungal infection.
107 Secondary syphilis: squamous syphilide of scrotum.
108 Secondary syphilis: circinate syphilide of penis.
109 Secondary syphilis: faint macular rash on face and trunk.
110 Secondary syphilis: typical papular syphilide of face. Note perioral distribution.

105

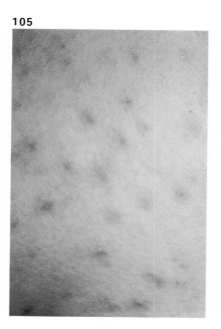

106

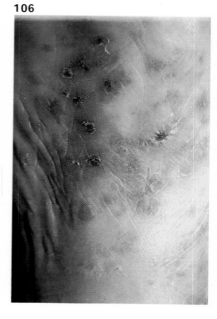

107

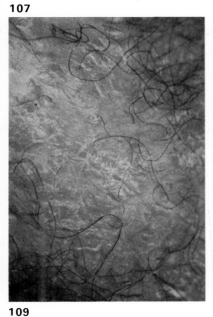

108

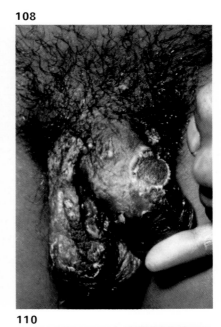

109

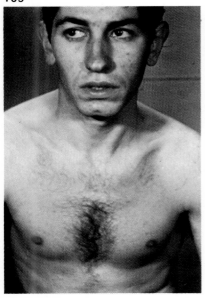

110

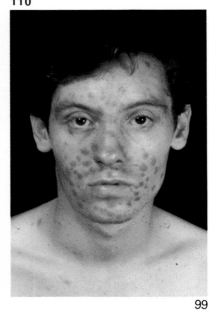

111 Secondary syphilis: maculo-papular syphilide. Note lesions of hair line ('corona veneris') and around mouth.
112 Secondary syphilis: papular syphilide. Note heaped-up (rupial) lesions on upper lip.
113 Secondary syphilis: papulo-squamous syphilide of face.
114 Secondary syphilis: papular syphilide of forehead and hair line.
115 Secondary syphilis: patchy alopecia ('moth-eaten scalp').

111

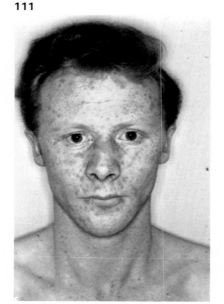

112

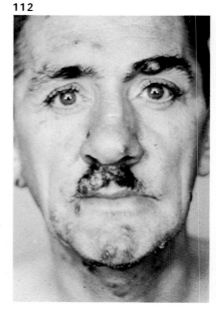

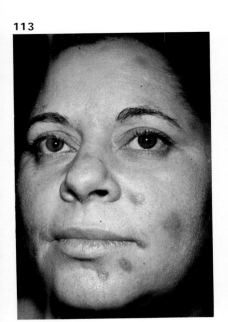

113

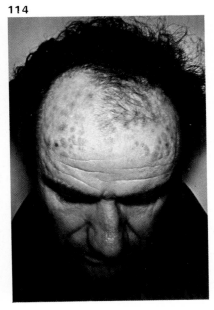

114

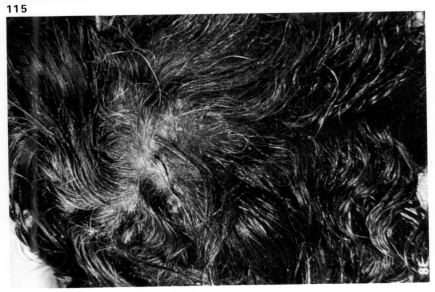

115

116 Secondary syphilis: patchy depigmentation of neck ('colli venus') and back after treatment.
117 Secondary syphilis: macular syphilide of chest.
118 Secondary syphilis: macular syphilide of abdomen and thigh.
119 Secondary syphilis: corymbose macular syphilide.
120 Secondary syphilis: well-marked corymbose lesions.

116

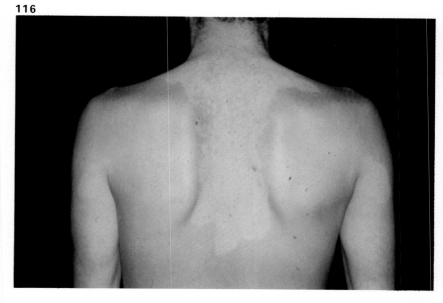

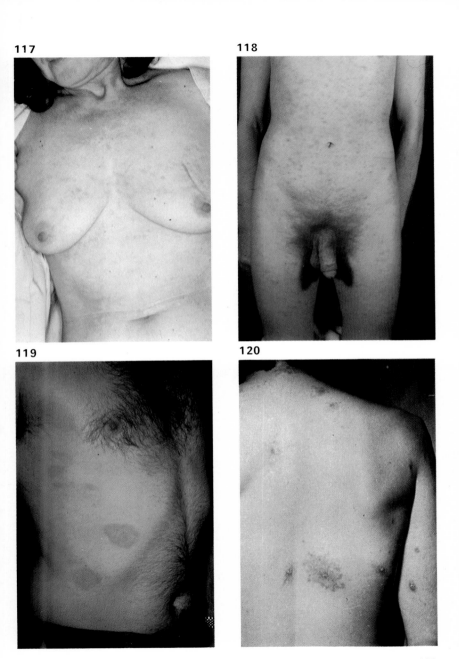

117

118

119

120

121 Secondary syphilis: faint macular syphilide of shoulders. The pink staining on the flanks is due to preparation for surgery.

122 Secondary syphilis: macular syphilide with some papular lesions.

123 Secondary syphilis: faint maculo-papular syphilide of back.

124 Secondary syphilis: maculo-papular syphilide of back.

125 & 126 Secondary syphilis: maculo-papular syphilide. The patient was subject to eczema and the rash was irritant.

121

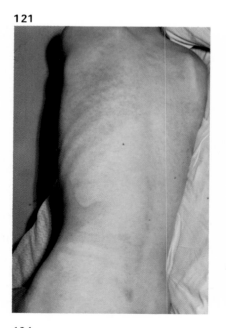

122

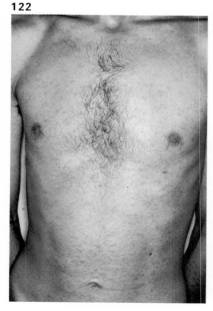

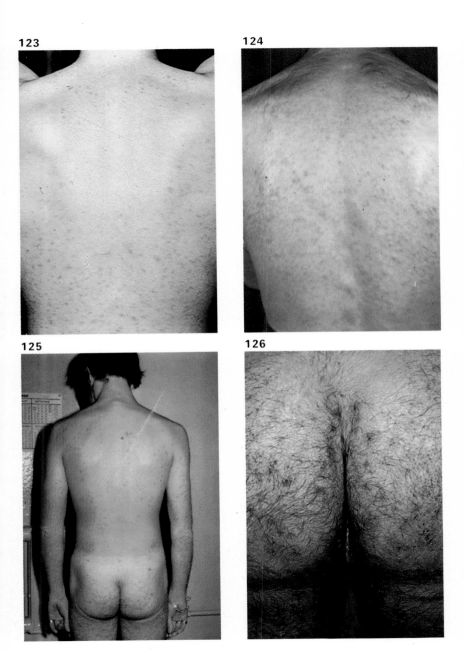

127 Secondary syphilis: maculo-papular syphilide of trunk.

128 Secondary syphilis: papulo-squamous syphilide: the adhesive plaster marks the site of skin biopsy.

129 Secondary syphilis: very typical papulo-squamous syphilide. Note the facial lesions, the colour and the symmetrical distribution. This rash had been present for three months before diagnosis and only cleared completely six months after completion of treatment.

130 Secondary syphilis: back of patient depicted in fig. 129. Note the distribution of the rash, reminiscent of pityriasis rosea (fig. 410).

131 Secondary syphilis: extensive maculo-papular syphilide.

132 Secondary syphilis: hyperpigmented papular syphilide. Lesions in Negro patients are often hyperpigmented.

127

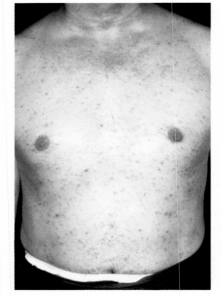

128

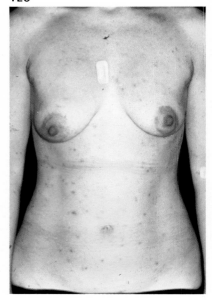

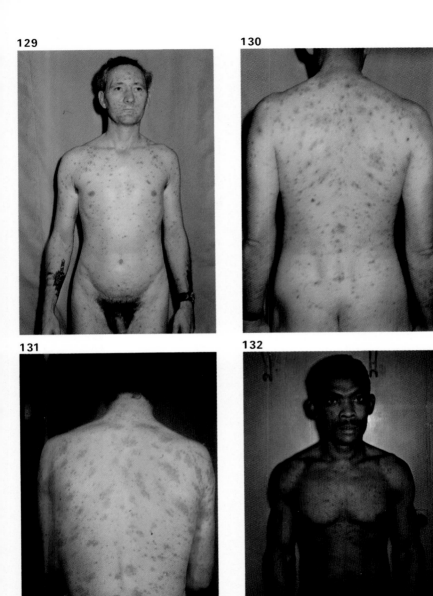

133 Secondary syphilis: papulo-squamous syphilide. Note palmar lesions.
134 Secondary syphilis: papulo-pustular syphilide.
135 Secondary syphilis: extensive palmar syphilide with marked squamous change.
136 & 137 Secondary syphilis: faint lesions of palmar and plantar syphilides.

133

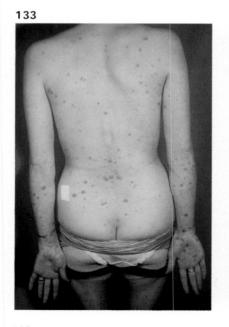

134

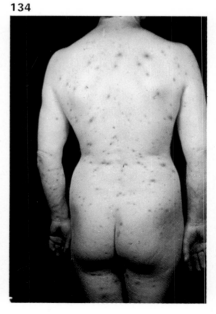

135

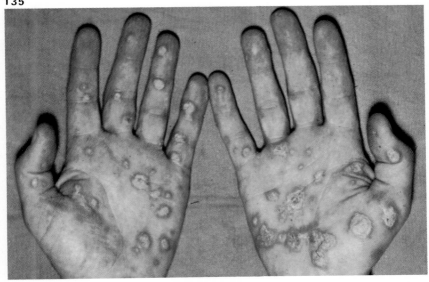

136

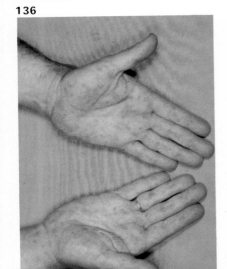

137

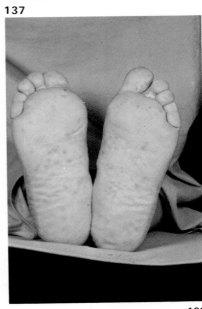

138 & 139 Secondary syphilis: palmar and plantar papulo-squamous syphilide.

140 Secondary syphilis: palmar and plantar syphilide (this patient is also depicted in fig. 106).

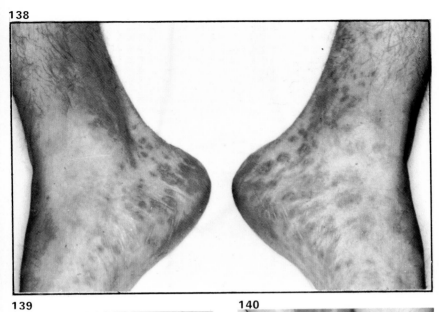

138

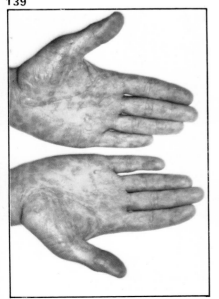

139

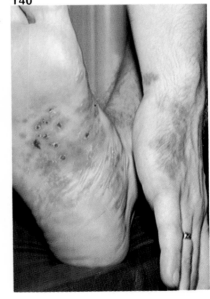

140

Condylomata lata

Condylomata lata are modified papular lesions found at anatomical sites where friction and moisture are present: they occur concurrently with the skin eruption. The regions where condylomata lata are most frequently found are the anus and vulva. Lesions may also occur on the penis, scrotum, thigh, axilla, angle of the mouth and beneath a pendulous breast. The lesions are pale brown or pale pinky-grey in colour, 5–20 mm in diameter. Initially· the lesions are discrete and circular but coalescence may occur to form a large lesion with a polycyclic outline. The surface of the lesion is sightly raised, flat and clean, and is usually moist from exuded serum. Micro-scopically this serum swarms with *T. pallida*. Condylomata lata are the most highly infectious lesions in syphilis.

141 Secondary syphilis: perianal condylomata lata. Note resemblance to warts (condylomata acuminata).

142 Secondary syphilis: mucous patches of scrotum and condylomata lata of thigh.

143 Secondary syphilis: condylomata lata of vulva.

144 Secondary syphilis: gross condylomata lata of vulva and anus.

141

142

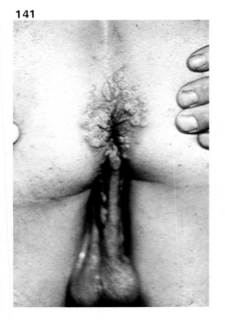
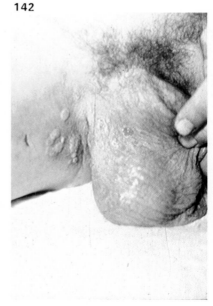

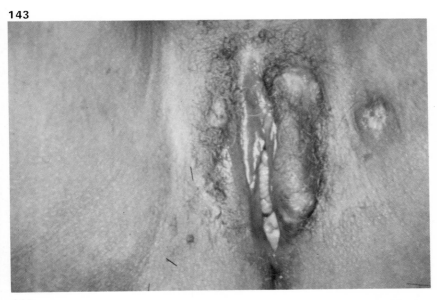

143

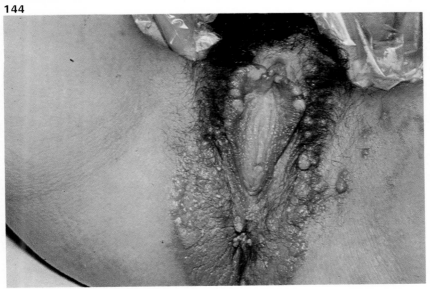

144

145 Secondary syphilis: gross condylomata lata of anus and buttock.
146 Secondary syphilis: condylomata lata of penis, mucous patches of scrotum and papulo-squamous syphilide of thigh.
147 Secondary syphilis: condylomata lata of vulva.
148 Secondary syphilis: condylomata lata of anus, acquired, in a 10 year old boy (Cf. yaws, fig. 223).
149 Secondary syphilis: condylomata lata of anus: small lesions.

145

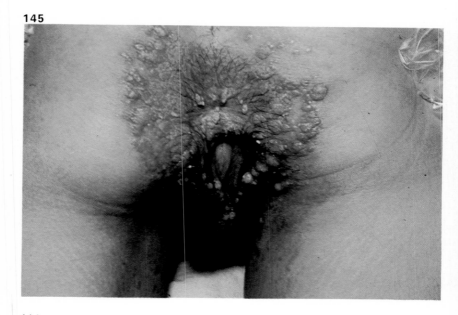

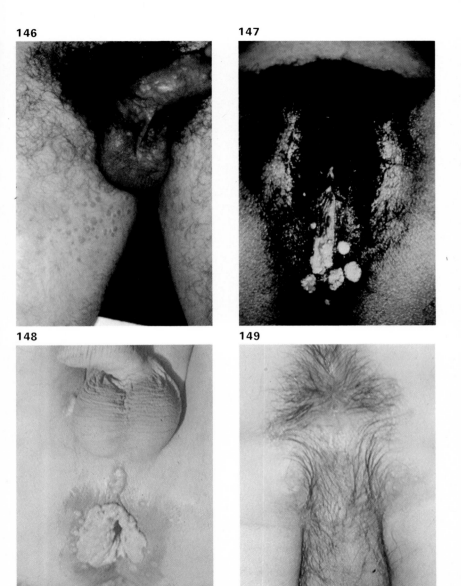

Lesions of mucous membranes

Lesions of the mucous membranes occur concurrently with the cutaneous eruptions in secondary syphilis. The lesions are termed *mucous patches*. Mucous patches are most commonly seen on the inner surface of the lips but are also found on other oral mucous membranes (including the tongue), in the pharynx and on the larynx. In the latter situation the lesion may cause hoarseness but the mucous patches are difficult to see without laryngoscopy. *Genital mucous membranes* may also be involved. In male patients mucous patches are found beneath the prepuce and on the glans penis; in female patients the lesions occur on the mucosal surfaces of the labia and vulva and, rarely, on the vaginal wall and cervix. All these lesions are highly infectious.

Mucous patches are round, oval or serpiginous (*'snail-track ulcer'*) in outline. The lesion is usually a superficial erosion but occasionally papules are seen. The mucous patch may be dull red in colour or covered with an easily removed grey membrane; the margin of the lesion is marked by an erythematous areola. On the surface of the tongue mucous patches look like bald areas: the appearance is due to destruction of the filiform papillae.

150 Secondary syphilis: mucous patches of prepuce. Note resemblance to herpes genitalis (Cf. fig. 506).
151 Secondary syphilis: mucous patches of prepuce.
152 Secondary syphilis: mucous patches of penis and prepuce.
153 Secondary syphilis: mucous patches of penis and prepuce.

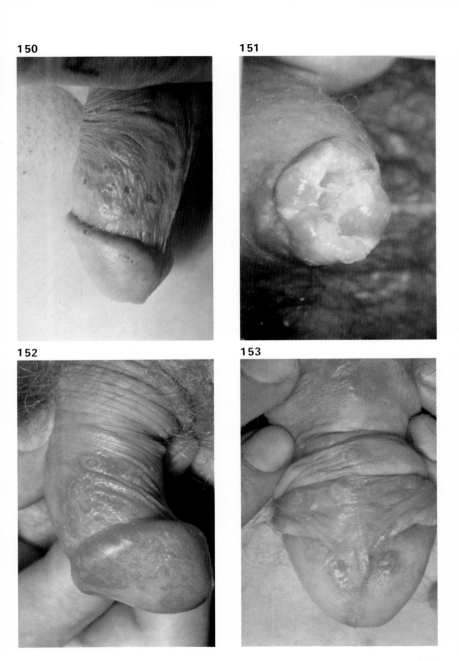

150

151

152

153

154 Secondary syphilis: mucous patches of penis. The lesions were irritant and originally thought to be due to scabies.

155 Secondary syphilis: mucous patches of scrotum (same patient as fig. 107).

156 Secondary syphilis: perianal mucous patches and maculo-papular syphilide.

157 Secondary syphilis: mucous patches of vulva, resembling herpes genitalis (Cf. fig. 514).

158 Secondary syphilis: mucous patches of upper thigh, resembling the perivulvitis often seen in trichomonal infestations (Cf. fig. 358).

159 Secondary syphilis: angular stomatitis and perioral rash. Note the 'split papule' at the angle of the mouth.

154

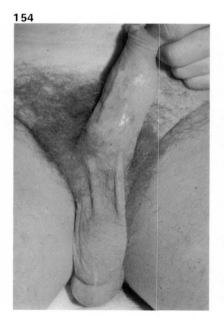

155

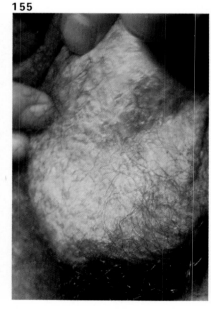

156

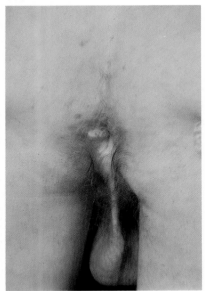

157

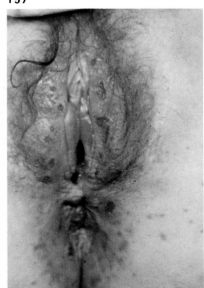

158

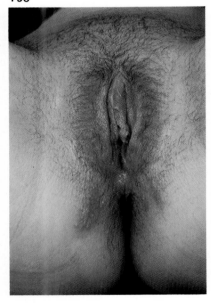

159

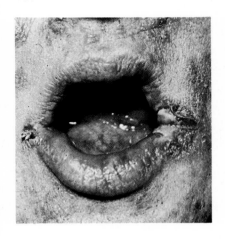

160 Secondary syphilis: 'snail track' ulcer.

161 Secondary syphilis: vesico-papular lesions and 'snail-track' ulcers of hard palate.

162 Secondary syphilis: mucous patch of upper lip with typical adherent exudate.

163 Secondary syphilis: papular and eroded lesions of hard palate.

164 Secondary syphilis: mucous patch of lower lip. Note resemblance to Behçet's syndrome (figs. 378 and 380).

160

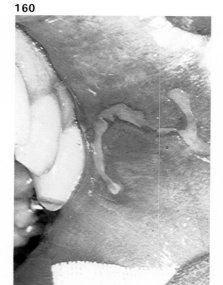

161

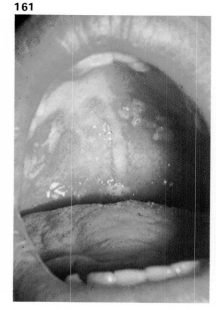

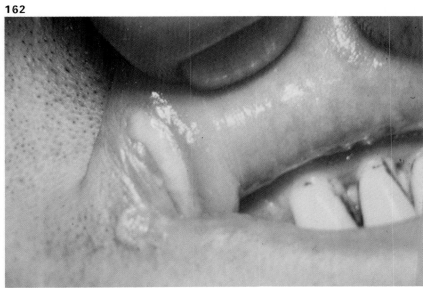

162

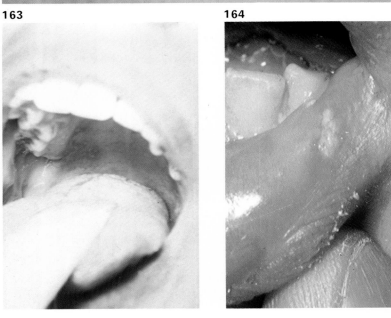

163

164

165 Secondary syphilis: mucous patches of upper lip: compare with erythema multiforme (fig. 551).

166 Secondary syphilis: mucous patches on tonsil: the pallid appearance is very typical (compare figs. 167 and 168).

167 Secondary syphilis: mucous patch of fauces. This lesion may be a primary chancre: corymbose skin lesions from the same patient are shown in fig. 120.

168 Secondary syphilis: lesion on soft palate. In my experience secondary syphilitic lesions of the tonsils, fauces, and pharynx are always pallid. These lesions may be difficult to distinguish from gummatous lesions of tertiary syphilis.

165

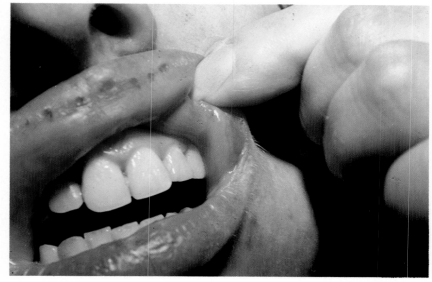

166

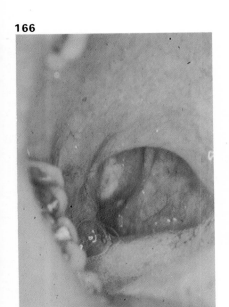

167

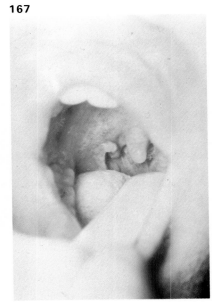

168

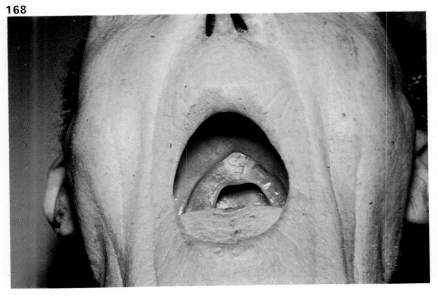

Latent syphilis

The stage in the progress of the infection known as *latent syphilis* begins when the superficial lesions and other manifestations of secondary syphilis resolve. The patient with latent syphilis has no outward signs of infection but positive serological tests (and sometimes changes in the C.S.F.) indicate the disease. Latent syphilis may persist for the rest of the life, but traditionally about 40 per cent of untreated latent syphilitics developed *tertiary syphilis*. My own experience suggests that the traditional ratio of 6 latent syphilis : 4 tertiary syphilis is no longer correct and that the ratio at present is nearer 10 : 1. The reason for this change may be that in the course of a lifetime many individuals with unrecognised syphilis receive for other conditions antibiotic therapy which is incidentally treponemicidal and which halts progression of the syphilitic infection.

Tertiary syphilis

At present, most patients found to have tertiary syphilis with the classical syndromes of *aortic regurgitation* and *aneurysm, tabes* and *general paralysis of the insane* will have initially attended and been diagnosed in other medical departments: they are later referred to the venereological clinic. Initial attendance of a patient with tertiary syphilis at a venereological clinic is therefore rare; for this reason a full account of the protean manifestations of tertiary syphilis is not given here. For fuller information readers are referred to comprehensive textbooks.

Summary

The basic pathological lesion in tertiary syphilis is a chronic granuloma known as a *gumma*. The gumma is an area of tissue necrosis, resulting from ischaemia due to endarteritis, surrounded by granulation tissue. It is a slowly progressive lesion; in most clinical cases areas of activity and regression can be recognised. The individual gumma may vary in size from 2–30 mm; often multiple lesions (gummata) are present. A diffuse fibrotic form of gummatous infiltration is occasionally seen.

The traditional figures show:

Approximately 15 per cent. Involvement of the skin, subcutaneous tissues, bones and periosteal tissues (*'benign late syphilis'*). Occurs 3–20 years after initial infection. Relatively more common in Negro patients.

Approximately 12 per cent. Involvement of the heart and/or aorta and (rarely) other parts of the cardio-vascular system. Occurs 10–30 years after initial infection. Relatively more common in males and Negroes.

Approximately 12 per cent. Involvement of the nervous system. The lesions may be meningeal, vascular or parenchymatous; mixed forms are common. Occurs 3–7 years (meningo-vascular) and 10–20 years (parenchymatous) after initial infection.

Other forms of tertiary syphilis are rare, although gummata have been described in practically every named anatomical structure. Some patients with no clinical abnormalities are found to have abnormal C.S.F. findings (asymptomatic neurosyphilis); the prognostic significance of this finding is not clear. Occasionally paroxysmal haemoglobinuria may occur in late syphilis due to the presence of a circulating haemolysin.

The features of tertiary syphilis

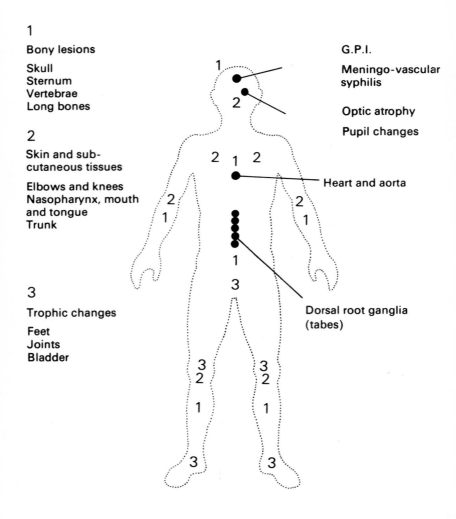

1

Bony lesions

Skull
Sternum
Vertebrae
Long bones

2

Skin and sub-cutaneous tissues

Elbows and knees
Nasopharynx, mouth
and tongue
Trunk

3

Trophic changes

Feet
Joints
Bladder

G.P.I.

Meningo-vascular
syphilis

Optic atrophy

Pupil changes

Heart and aorta

Dorsal root ganglia
(tabes)

Skin lesions

Nodulo-cutaneous syphilides Nodules, single or grouped, appear in the skin and slowly increase in size. The whole lesion may be extensive but individual nodules seldom exceed 1 cm in diameter. The lesion may remain nodular or, more commonly, ulcerates. Most lesions show evidence of simultaneous activity and regression, with ulceration at the periphery and healing at the centre of the involved area. The ulcer is *'punched-out'* and is circular (or polycyclic) in outline; groups of lesions may show annular, gyrate or serpiginous topography. There is often an adherent basal slough, termed *'wash-leather slough'*. Squamous change (scaling) may overlie non-ulcerated nodules occurring on the palms and soles. After treatment or spontaneous healing characteristic non-contractile *'tissue-paper'* scarring persists, often showing areas of hypo- or hyper-pigmentation.

Gummatous ulcers are painless and may cause severe tissue destruction before the patient seeks advice. They are most frequently found on the face, trunk and thigh but no region is exempt.

Subcutaneous gummata These gummata originate as subcutaneous or deeper (*e.g.* periosteal) nodules which become attached to and later ulcerate through the skin. The lesions are usually single and may be up to 4–5 cm in diameter. The clinical appearance and location is similar to the nodulo-cutaneous syphilides. Severe and deep tissue destruction may occur.

169 Tertiary syphilis: nodulo-cutaneous ulceration of chest wall, showing peripheral activity, central healing and hypopigmentation.
170 Tertiary syphilis: annular lesion of arm.
171 Tertiary syphilis: nodulo-cutaneous syphilide of back with hyper-pigmented scarring.
172 Tertiary syphilis: multiple gummatous ulcers of chest wall.
173 Tertiary syphilis: typical gummatous ulcers, showing vertical walls and 'wash-leather' slough.

169

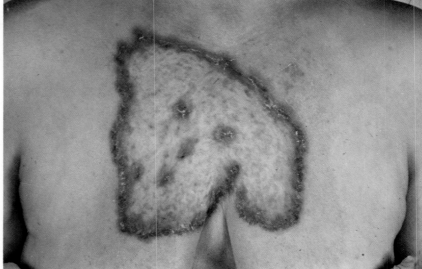

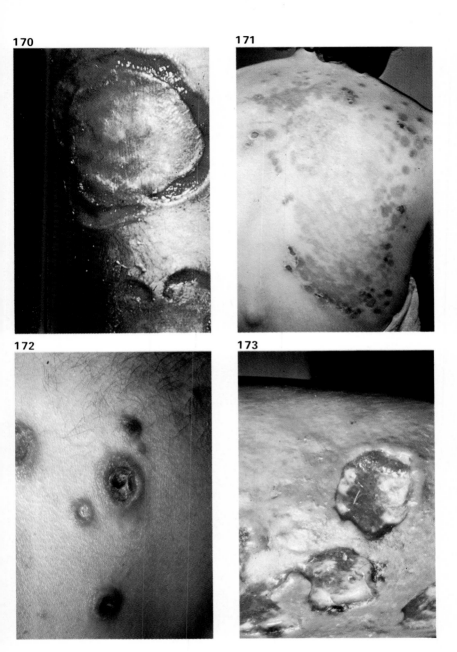

174 Tertiary syphilis: hypopigmented 'tissue-paper' scarring at sites of healed gummatous ulcers.
175 Tertiary syphilis: hypopigmentation following healing of extensive gummatous ulceration. Compare with fig. 214 (late pinta) and fig. 225 (yaws of foot).
176 Tertiary syphilis: gumma of penis.
177 Tertiary syphilis: gummatous ulcers of calf.
178 Tertiary syphilis: 'tissue-paper' scarring over healed gumma of hip.

174

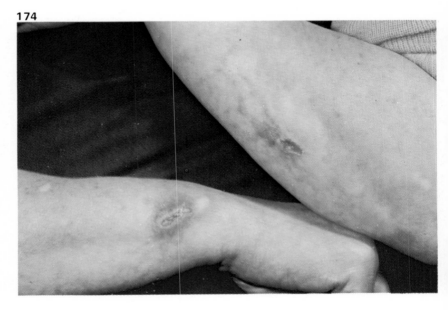

175

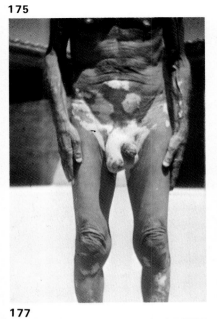

176

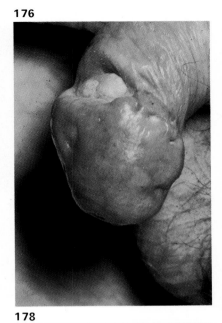

177

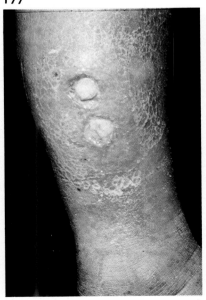

178

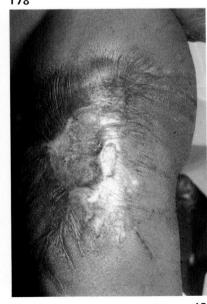

Cardio-vascular lesions

It has not been thought appropriate here to describe in detail late
syphilis of the cardio-vascular system. Patients with these conditions are
rare and usually present at cardiac, thoracic or vascular clinics. The
classical syndromes are:

Aortic regurgitation
Aortic aneurysm   Mixed forms are
Coronary ostial stenosis   frequently encountered

Syphilitic aortitis Precedes the development of the recognisable syndrome
but probably cannot be diagnosed *ante-mortem*. Occasionally syphilitic
involvement of other arteries may occur. About 30 per cent of patients with
cardio-vascular syphilis also have symptomatic or asymptomatic neurosyphilis. Examination must always include neurological assessment and
lumbar puncture.

179 Tertiary syphilis: aneurysm of ascending aorta which has eroded
sternum and extended onto chest wall.
180 Tertiary syphilis: X-ray of aortic aneurysm. Note linear calcification in
aortic wall.

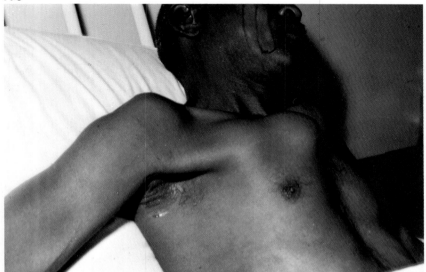

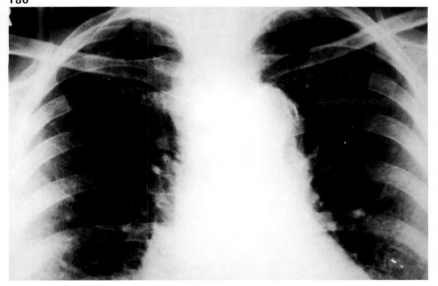

Neurosyphilis

As with cardio-vascular syphilis it has not been thought appropriate
to describe the now rare syndromes of neurosyphilis in detail. The classical
syndromes are:

General paralysis of the insane (G.P.I.) ⎤ Mixed forms are
Tabes dorsalis (Locomotor ataxia) ⎦ common (*Taboparesis*)

The meninges and vascular supply of the brain and spinal cord may be
involved; occasionally *primary optic atrophy* may be the sole abnormality
in neurosyphilis. *Asymptomatic neurosyphilis* (abnormal C.S.F. findings
with normal neurological examination) is nearly as frequent as symptomatic
neurosyphilis. Cardio-vascular syphilis may co-exist with neurological
involvement. Complete examination must include clinical and radiological
assessment of the heart and aorta.

181 Tertiary syphilis: primary optic atrophy.
182 Tertiary syphilis: perforating ulcers of foot in tabes dorsalis.
183 Tertiary syphilis: Charcot arthropathy of knees: tabes dorsalis.
184 Tertiary syphilis: Charcot arthropathy of elbow: tabes dorsalis.

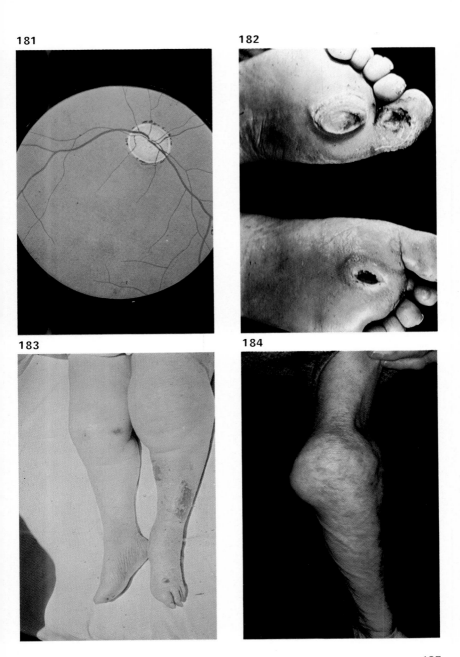

181

182

183

184

185 Tertiary syphilis: Charcot arthropathy of knees: tabes dorsalis.
Note the hypermobility of the joint.
186 Tertiary syphilis: X-ray of fig. 185.

185

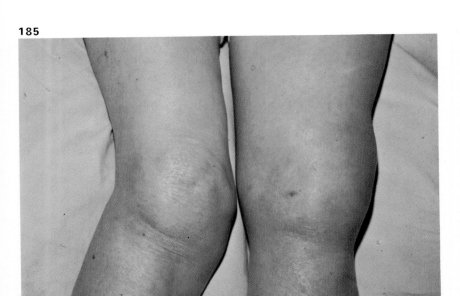

186

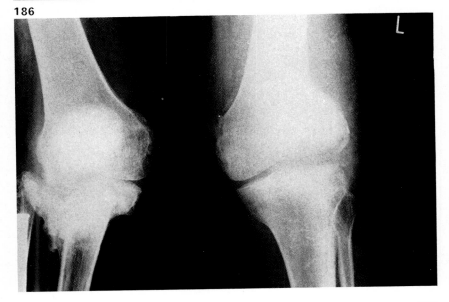

Bone lesions

Tertiary syphilis of the bones generally begins as a *periostitis,* with later cortical involvement. The osteo-periostitis of long bones is predominantly an osteoblastic process (there is usually radiological evidence of adjacent osteoclastosis) with deposition of extra bone under the periosteum, resulting in the usual clinical findings of thickening and irregularity of affected areas. Syphilitic osteo-periostitis of the bones of the skull and nasal skeleton is predominantly osteoclastic, giving the clinical finding of bone loss, which occasionally is very extensive.

About 50 per cent of patients with tertiary syphilis of bone complain of pain in affected areas; this pain may be very severe and is classically described as 'boring' in character. Even after anti-syphilitic treatment the pain may persist for a long time. Tenderness of the affected areas is often found. Radiological survey of patients with symptomatic bony syphilis often discloses further areas of asymptomatic involvement. Pathological fractures may occur and are sometimes the presenting symptom.

187 Tertiary syphilis: X-ray of osteo-periostitis of skull. Note areas of osteoclastosis and osteoblastosis.
188 Tertiary syphilis: X-ray of osteitis of humerus.
189 Tertiary syphilis: X-ray of periostitis of tibia. Compare with 'sabre' tibia in yaws, fig. 228.

187

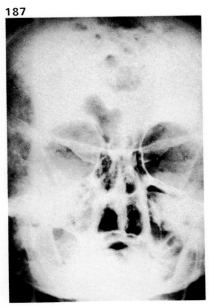

188

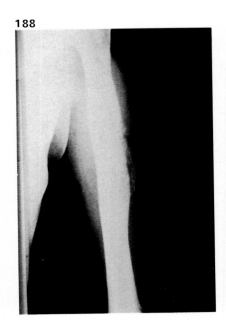

189

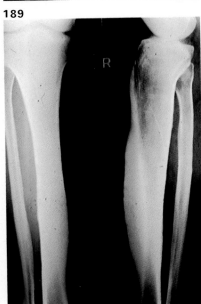

Oral cavity lesions

The oral and nasal cavities are commonly a site of involvement in
tertiary syphilis. Gummatous ulceration may be found on any of the
mucous membranes. Diffuse gummatous infiltration of the tongue may result
in chronic superficial or interstitial glossitis: rarely diffuse infiltration may
involve the lips. Syphilitic osteitis and periostitis may involve the bones of
the nose or palate; palatal perforation and destruction, sometimes gross,
of the nasal skeleton may be found.

190 Tertiary syphilis: chronic interstitial glossitis and leukoplakia.
191 Tertiary syphilis: chronic interstitial glossitis. This lesion was completely
asymptomatic and persisted after treatment.
192 Tertiary syphilis: periosteal gummata of nasal bones.
193 Tertiary syphilis: gross gummatous destruction of face. Compare with
figs. 208 (endemic syphilis) and 232 (yaws).

190

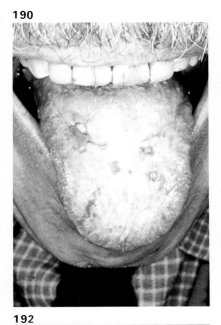

191

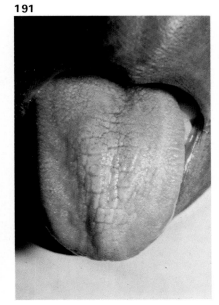

192

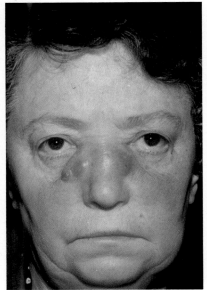

193

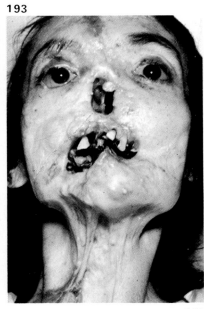

Congenital syphilis

In untreated syphilis it is thought that waves of spirochaetaemia occur, diminishing in frequency as time passes. Transmission of syphilis to the foetus occurs when pregnancy coincides with spirochaetaemia. The observed great variation in severity of congenital syphilis is probably dependent on the degree of spirochaetaemia at the time the foetus is initially infected.

In acquired syphilis the portal of entry of infection is the site of the primary chancre; in congenital syphilis *Treponema pallida* enter the circulation directly through the placental capillaries. The clinical manifestations correspond to the secondary and later stages of acquired syphilis.

Congenital syphilis is a rarity in developed countries, where ante-natal care includes serological examination for syphilis so that affected mothers are treated before delivery. In these circumstances active congenital syphilis is almost unknown but children infected *in utero* may sometimes show clinical evidence of the effects of the infection present (*e.g.* Hutchinson's teeth) before treatment was begun.

Detailed consideration of congenital syphilis is given in the standard textbooks. In this Atlas only a few representative pictures of clinical manifestations are included. It is very important to remember that a diagnosis of congenital syphilis is the starting point for a full family investigation.

194 Congenital syphilis: facies after plastic surgery. Note bridge of nose.
195 Congenital syphilis: scarring resulting from extensive gummata.
196 Congenital syphilis: profile of facies. Note the depressed bridge of the nose.
197 Congenital syphilis: facies. Note scarring overlying mandible.

194

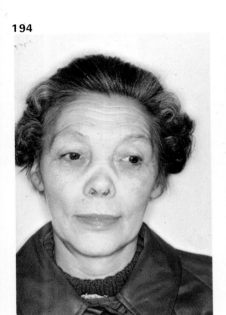

195

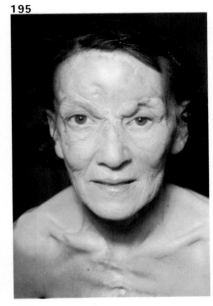

196

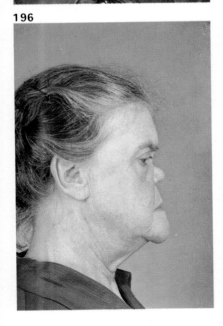

197

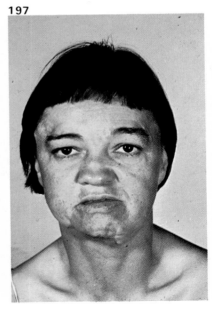

198 Congenital syphilis: Hutchinson's incisors: the 'screwdriver' shape with marginal notching and distal narrowing.
199 Congenital syphilis: perioral rhagades, scarring from healed perioral lesions. Similar lesions are sometimes seen in the perianal region.
200 Congenital syphilis: faint corneal nebulae resulting from previous interstitial keratitis.

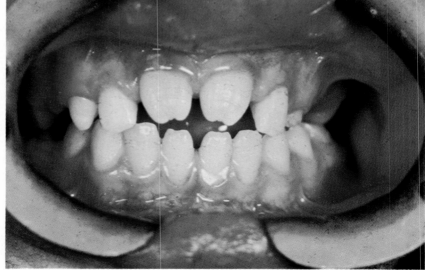

199

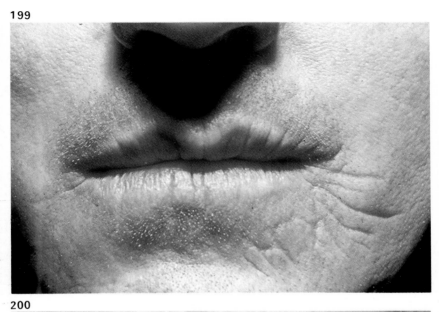

200

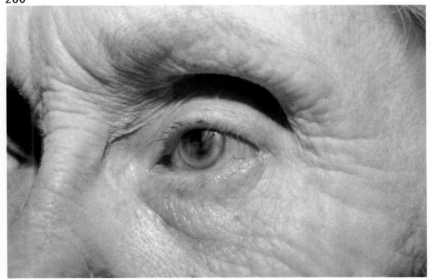

201 Congenital syphilis: further example of faint corneal nebulae.
202 & 203 Congenital syphilis: extensive scarring from previous gummata: same patient as fig. 195.
204 Congenital syphilis: perforation of hard palate.
205 Congenital syphilis: choroido-retinitis.

201

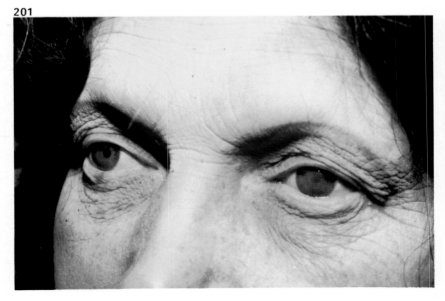

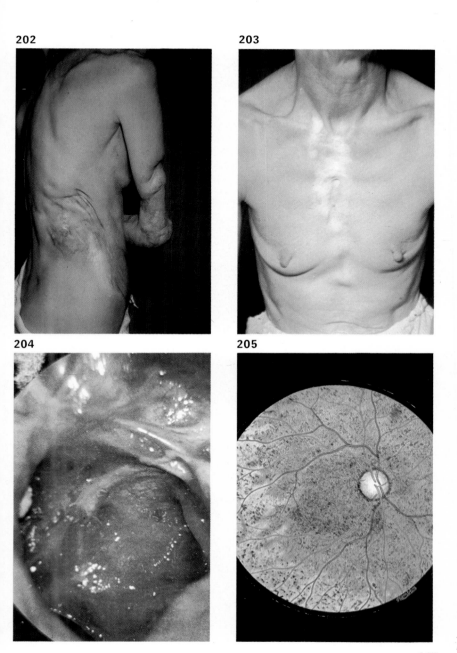

NON-VENEREAL TREPONEMATOSES

Endemic syphilis

Endemic syphilis is the non-venereal form of the disease. Most infections are acquired in childhood with subsequent diminished susceptibility to sexually transmitted treponematosis in adult life. It is a disease found in communities with low socio-economic status and was formerly widespread. With rising standards of nutrition, clothing and hygiene the incidence of endemic syphilis declines; at present, foci of infection are still to be found in some areas of Africa, the Middle East and Australia. Endemic syphilis is essentially the same disease throughout the world but each area has its own local name for the disease. Extant names include *bejel* (Iraq) and *njovera* (Rhodesia) inter alia; historical names include the *button scurvy* (Ireland) and *skerljevo* (Balkan countries).

The infectious stages of the disease are mainly seen in children, as the disease is spread by direct contact (usually to other children), or by shared objects such as drinking or eating utensils and, possibly, by insect vectors.

Clinically the disease resembles venereal syphilis but primary lesions are rarely identified. Many patients present with moist lesions of the oral mucous membranes or condylomata lata, or may be found to have latent infection or skin or bone gummata. Cardio-vascular and neurological involvement seems to be rare. Congenital transmission is also rare as untreated females are generally non-infectious when the fertile years are reached; conversely an infant infected from another source may super-infect the parent.

Treponema pallida seen in material in lesions of endemic syphilis are indistinguishable from those seen in venereal syphilis. The distinction between the two forms of the disease is not absolute but depends on consideration of clinical and epidemiological factors.

206 Early endemic syphilis: annular rash on penis. Compare with fig. 108.
207 Early endemic syphilis: anal papillomata: compare with condylomata lata figs. 141, 148 and yaws, fig. 223.
208 Late endemic syphilis: gangosa (gummatous destruction of the face). Compare with figs. 196 (congenital syphilis) and 232 (yaws).
209 Late endemic syphilis: destruction due to nasal gummata in bejel.

206

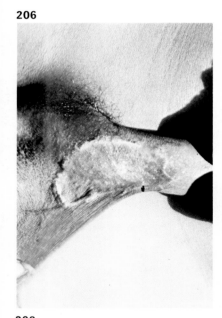

207

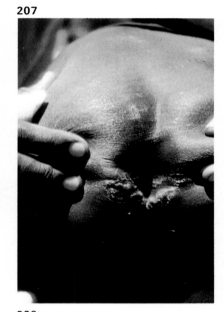

208

209

Pinta (mal de pinto, carate)

Pinta is the least serious of the treponematoses. The disease has a limited distribution (see facing map) and is almost confined to poor native and Negro communities. Transmission is by contagion and nearly always occurs before adult life.

The clinical lesions are known as *pintides*. The *primary pintide* appears, usually on exposed skin surfaces, after an average incubation period of six to eight weeks. The lesion is circular, erythematous and desquamating; the diameter is 1–2 cm. *Secondary pintides* appear several months later. These are of similar form and may be adjacent to the primary lesion or disseminated. Hyperpigmentation of secondary lesions is common; colour may be reddish, bluish or black. *Tertiary pintides* appear after a latent period of some years; dyschromia is usual, with late lesions often completely depigmented. Hyperkeratosis may occur in secondary and tertiary stages and is particularly common with pintides on the palms or soles. Visceral involvement and congenital transmission (if they occur at all) are extremely rare.

The causative organism is *Treponema carateum*: it may be recognised in dark-ground specimens taken from primary and secondary pintides. *T. carateum* is morphologically identical with *T. pallidum*. Serological tests for syphilis give positive results in pinta, indistinguishable from the results obtained in venereal syphilis.

210 Pinta: distribution.
211 Pinta: early pintide. Note the similarity to maculo-papular syphilides.

210

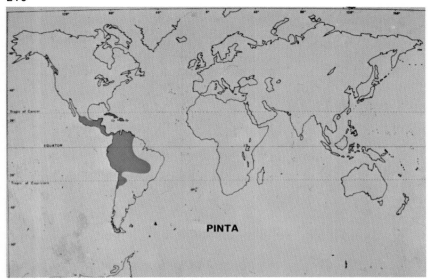

PINTA

211

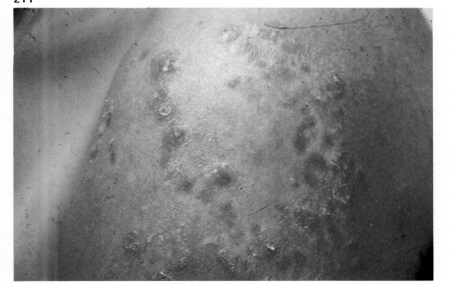

212 Pinta: pintide of nose and cheek showing typical slate-blue pigmentation.
213 Pinta: late pintide of leg: hyperpigmented lesions.
214 Pinta: late pintide: 'glove hand' showing achromia.

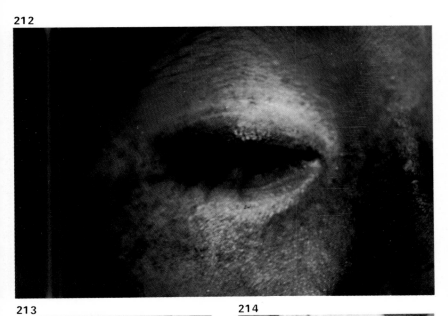

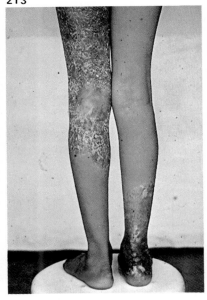

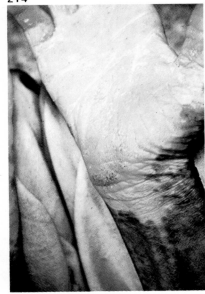

Yaws (pian, framboesia)

Yaws, the non-venereal treponematosis of tropical regions, was formerly extremely common. Mass treatment campaigns organised by the World Health Organisation have greatly reduced the incidence. Transmission of the disease is by contagion and occurs in most cases in childhood. The pattern of the disease is broadly similar to syphilis, as primary, secondary, latent and tertiary stages occur and clinical lesions show considerable likeness. As in the other non-venereal treponemal diseases, cardio-vascular and neurological involvement and congenital transmission are thought to be very rare.

The *primary* lesion (*mother yaw*) may be found anywhere on the skin but is usually on an exposed surface such as the lower part of the leg. The lesion is initially a papule which enlarges to an exuberant granuloma with papillomatous margins, 2—6 cm in diameter. Local lymph gland enlargement is common. The mother yaw is often secondarily infected with pyogenic organisms so that residual scarring is frequent.

In *secondary yaws* skin eruptions similar to those seen in secondary syphilis but often hypertrophic are found. The most characteristic eruption is the papillomatous rash—*framboesioma.* The multiple red (or yellow crusted) papillomata, 5—20 mm in diameter, are found mainly at muco-cutaneous junctions and on other moist surfaces of the body, but may occur on any skin surface. Painful papillomata and hyperkeratosis may occur on the soles; the term *'crab yaws'* has been applied to these lesions which cause the patient to walk with a crab-like gait. Generalised lymph gland enlargement is often found. Painful areas of osteitis and periostitis may occur: dactylitis is seen in children; characteristic broadening of the bridge of the nose due to maxillary periostitis is known as *goundou.*

In *tertiary yaws* skin lesions similar to nodulo-cutaneous syphilides and gummatous ulcers occur and may later heal with extensive scarring and hypopigmentation. Painful bone lesions are often found, particularly affecting the long bones. The characteristic lesion is the *'sabre tibia'* Extensive bone- and soft-tissue involvement of the face sometimes occurs; the condition is known as *gangosa* and may result in gross deformity. Juxta-articular nodes are another characteristic lesion of late yaws.

The causative organism, *Treponema pertenue,* can be found in dark-ground preparations taken in the primary and secondary stages from skin lesions and enlarged glands. It is morphologically identical with *Treponema pallidum.* The serological tests for syphilis give positive results in yaws. A recurrent problem is the evaluation of positive serological tests for

215 Yaws: distribution.
216 Yaws. Primary yaws. Compare with primary syphilis, figs. 57 and 61.

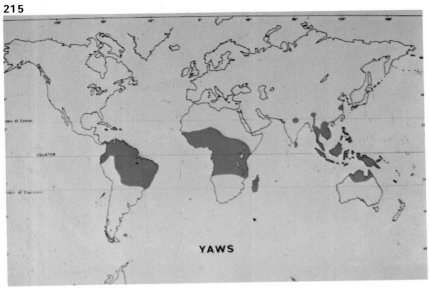

215

YAWS

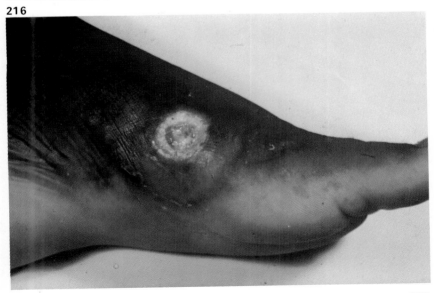

216

syphilis in clinically normal patients who originate in yaws-endemic areas:
a definite diagnosis is usually impossible to establish.

217 Yaws. Primary yaws: hypertropic lesion.
218 Yaws. Secondary yaws: framboesioma.
219 Yaws. Secondary yaws: hypertropic papulo-squamous skin lesions.
220 Yaws. Secondary yaws: facial papillomata.

217

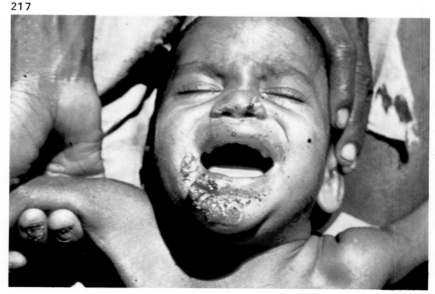

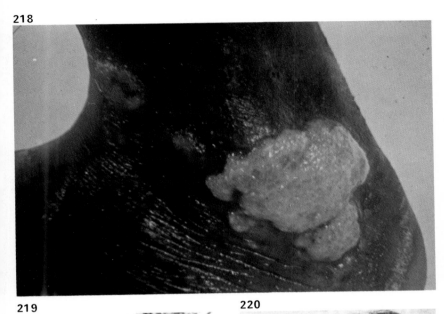

218

219

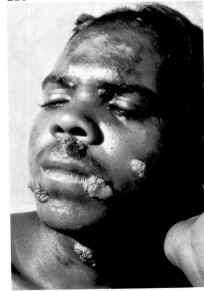

220

221 Yaws. Secondary yaws: split papule at corner of mouth. Note flies, possibly a vector of infection.
222 Yaws. Secondary yaws: condylomata and papillomata.
223 Yaws. Secondary yaws: condylomata of anus. Compare with fig. 141 (secondary syphilis).

221

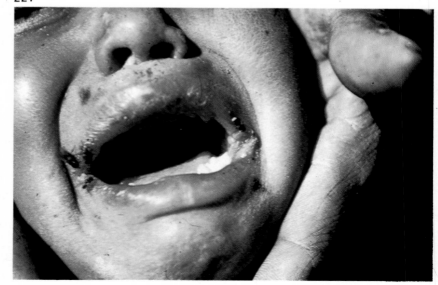

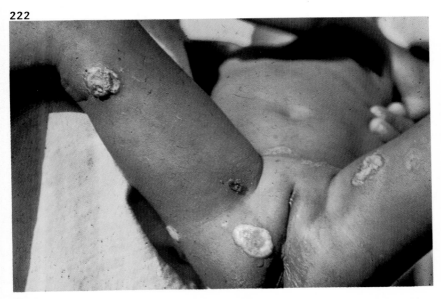

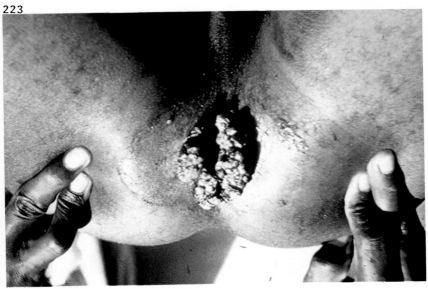

224 Yaws. Secondary yaws: split papules of lips.
225 Yaws. Secondary yaws: 'crab yaws' of foot; note 'worm-eaten' appearance.
226 Yaws. Secondary yaws: corymbose lesions. Compare with fig. 120.
227 Yaws. Late yaws: gummata of arm. Compare with fig. 202.
228 Yaws. Late yaws: 'sabre tibiae', a very typical lesion. The hand belongs to a nurse.

224

225

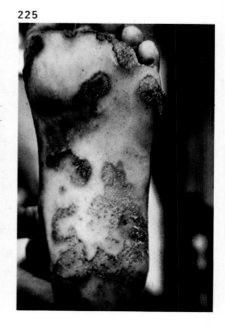

226

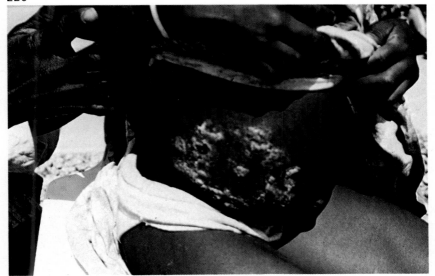

227

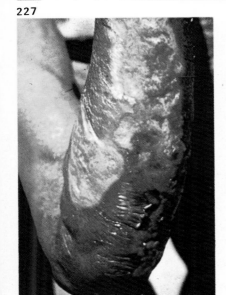

228

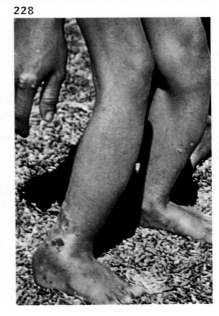

229 Yaws. Late yaws: annular gummatous lesions.
230 Yaws. Late yaws: extensive gummata, shortly after treatment was commenced.
231 Yaws. Late yaws: early gangosa.
232 Yaws. Late yaws: late gangosa.

229

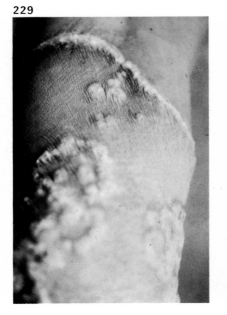

230

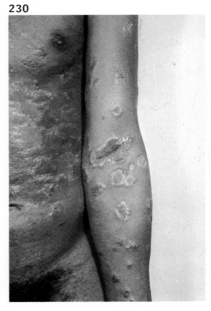

231

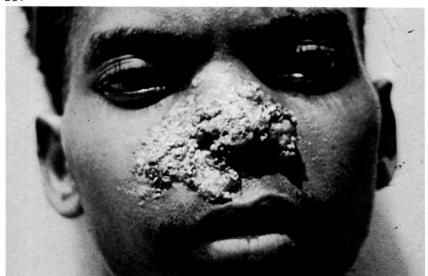

232

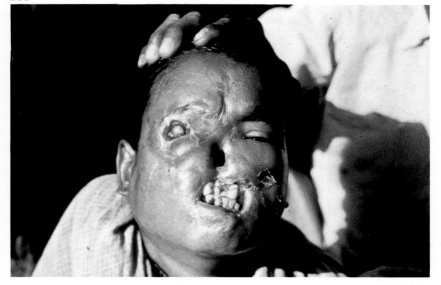

CHANCROID, GRANULOMA INGUINALE & LYMPHOGRANULOMA VENEREUM

Chancroid (soft sore, ulcus molle, soft chancre)

Chancroid is an acute sexually transmitted infection of the genitalia characterised by painful ulceration and frequent bubo formation. It is found throughout the world but is much more frequently seen in warm climates and in populations with low standards of hygiene; in Western Europe and North America the disease is now rare. Chancroid is highly infectious but clinical lesions are disproportionately rare in women suggesting that a 'carrier' state occurs.

Cause
The causative organism is a bacterium, *Haemophilus ducreyi.*

Clinical course
The incubation period averages three to six days but shorter and longer periods have been reported occasionally. The clinical lesion begins as a painful vesicular papule which rapidly progresses to an ulcer with a bright red areola and shelving margins. The ulcer may be round or irregular in outline; adjacent lesions often become confluent. Secondary infection is common and considerable tissue destruction may ensue. Lesions may be single or multiple, with diameter ranging from 3–20 mm. Chancroidal ulcers may be found anywhere on the genitals but are most frequent at the sites where trauma during intercourse is most likely: in males at the preputial margin or on the coronal sulcus and fraenum; in females on the labia and perineum. Occasionally extra-genital lesions, usually due to auto-inoculation from adjacent genital ulcers, are found.

Painful enlargement of the inguinal lymph nodes (*chancroidal bubo*) occurs 7–14 days later in about 50 per cent of cases. Sometimes the disease presents with the bubo, the initial ulcer having been unnoticed. The enlarged glands may be uni- or bi-lateral and usually lie above the inguinal ligament. The bubo is a mass of glands matted together and is often adherent to the overlying erythematous and oedematous skin. Central softening is often found, and, untreated, the bubo may rupture and discharge through a fistula. Occasionally extensive involvement of the skin around the fistulous opening occurs.

233 *Haemophilus ducreyi*. Photomicrograph showing 'school of fish' growth pattern.
234 *Haemophilus ducreyi.* Photomicrograph of smear from lesion. Short Gram-negative rods, singly or in chains, can be seen. Pappenheim stain.
235 Chancroidal ulcer of fraenum.
236 Chancroidal ulcers of coronal sulcus. Note irregularity, shelving edges and secondary infection.

233

234

235

236

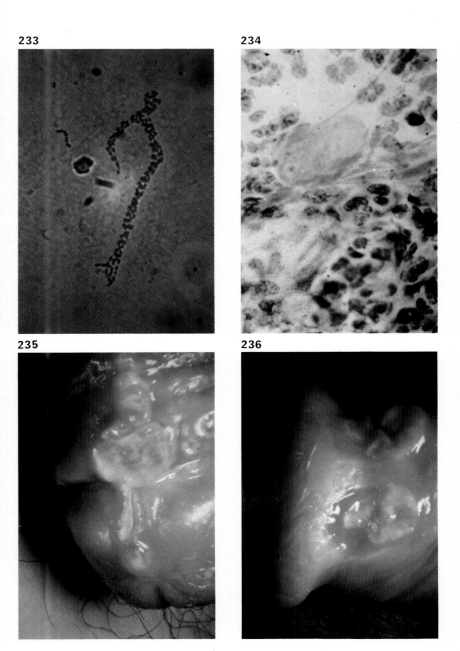

Diagnosis

Many cases are diagnosed on clinical grounds alone; it is important to remember that concurrent syphilis may be present.

Haemophilus ducreyi may be recognised in stained preparations made from the surface of the ulcer or bubo pus, but recognition is often difficult because of other organisms present. Culture is difficult.

Positive results with the Dmelcos and Ito—Reenstierna skin tests should be assessed cautiously because the positivity may be a permanent change residual from previous infection.

237 Chancroid of prepuce: very typical lesion.
238 Another example of chancroid of prepuce.
239 Chancroid of penis and inguinal bubo.

237

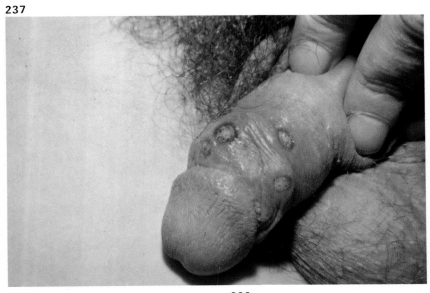

238

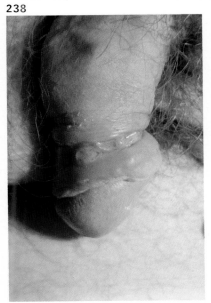

239

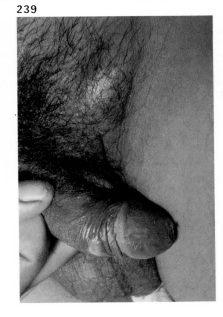

167

CHANCROID, GRANULOMA INGUINALE
& LYMPHOGRANULOMA VENEREUM

240 Chancroid of vulva.
241 Typical chancroid of glans penis. Note considerable tissue destruction and adherent slough.
242 Chancroid of prepuce. Note the considerable oedema and erythematous halo. After treatment of this type of lesion phimosis may occur.
243 Penile chancroid.
244 Penile chancroid with secondary bacterial infection.

240

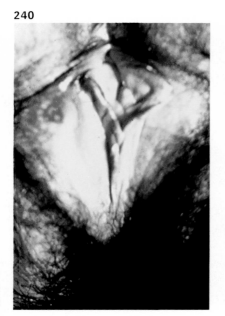

241

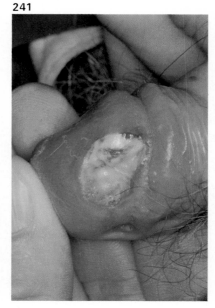

242

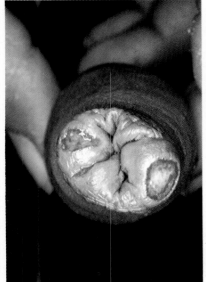

243

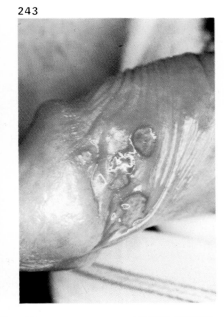

244

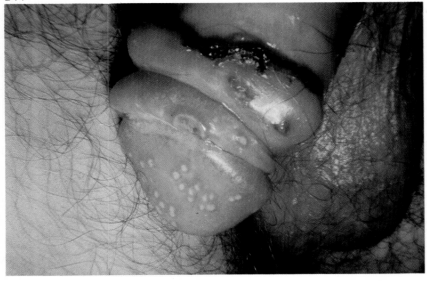

Granuloma inguinale (granuloma venereum, Donovanosis, granuloma contagiosa)

Granuloma inguinale is a chronic, slowly progressive, ulcerative disease of the genitalia and adjacent tissues. It is probable that transmission is by sexual contact although this has not yet been definitely proved. Sexual consorts of patients with the disease are often uninfected. The disease is most often seen in Negro or other coloured races, and is most frequently encountered in tropical or sub-tropical areas. Males are more frequently affected than females.

Cause
The causative organism is an intracellular parasite, *Donovania granulomatosis*. The organism was first recognised in 1905.

Clinical course.
The incubation period is usually between 10 and 40 days, but periods as short as three days and as long as 84 days have been reported.

The clinical lesion begins as a painless vesicle or indurated papule. This lesion becomes eroded to form an ulcer with a beefy red granular base. The ulcer is usually round with a rolled edge. Centrifugal spread occurs, usually eccentrically, and the advancing edge of the lesion spreads on the surface from the primary site to adjacent tissue. Subcutaneous extension also occurs—in the inguinal region this may be mistaken for lymph gland involvement and is known as a *pseudo-bubo*. Subcutaneous abscesses may occur. In the absence of treatment healing is uncommon and the granulated area may become secondarily infected or the site of neoplastic change.

Lesions of granuloma inguinale are found initially on the shaft of the penis, the labia or the perianal region and (rarely) on the vaginal wall or cervix. Spread of the disease commonly involves the groin, perineum or natal cleft. Rarely, extra-genital lesions may be found.

Diagnosis
Donovania granulomatosis can usually be easily identified in suitably stained smears of material taken from the active edge of a lesion. Culture is possible but difficult. The painless ulceration in early stages may mimic syphilis, and tests to exclude syphilis are essential.

245 Photomicrograph of tissue smear showing *Donovania granulomatis*. Note the deeper staining at the poles of the organism ('safety pin').

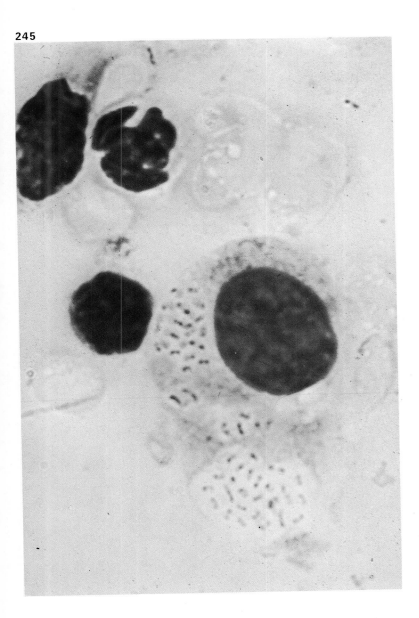

CHANCROID, GRANULOMA INGUINALE
& LYMPHOGRANULOMA VENEREUM

246 Early lesion of prepuce.
247 Lesion on penile shaft.
248 Early ulcerative lesion of coronal sulcus.
249 Lesion on penile shaft: uneven extension of infection has made the outline irregular.
250 Exuberant lesion of prepuce.

246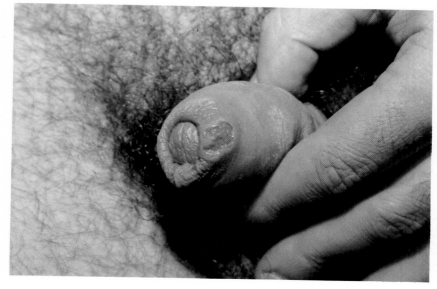

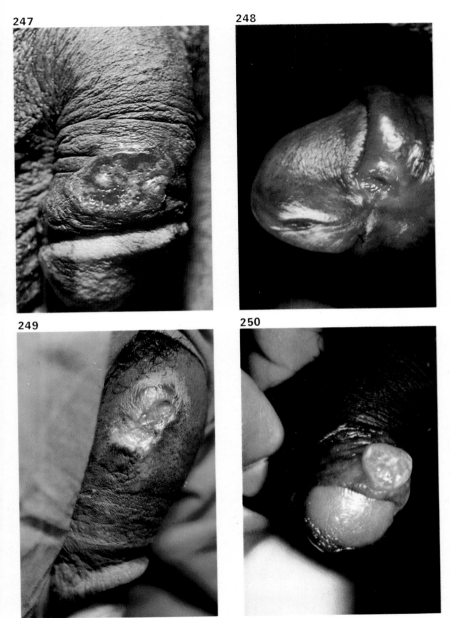

CHANCROID, GRANULOMA INGUINALE
& LYMPHOGRANULOMA VENEREUM

251 Lesion on penile shaft showing 'rolled edge'.
252 Late granuloma inguinale involving vulva and natal cleft.
253 Late granuloma inguinale: extensive involvement of scrotum and pathological amputation of the penis.
254 Late granuloma inguinale: involvement of penis and scrotum with gross scarring and contraction.
255 Late granuloma inguinale of groin.

251

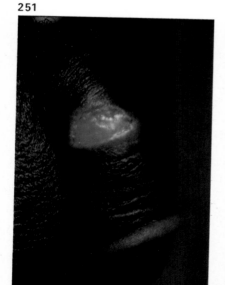

252

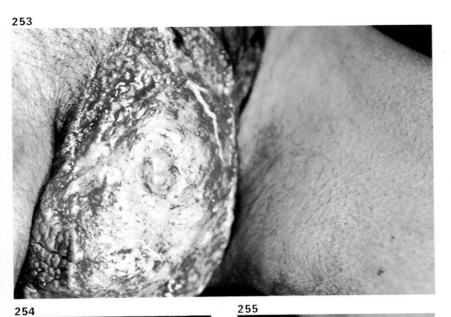

253

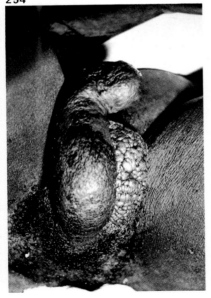

254

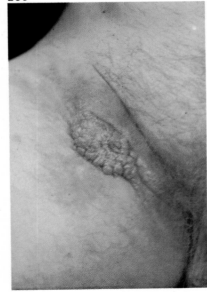

255

Lymphogranuloma venereum (lymphogranuloma inguinale, Nicolas–Favré disease, climatic bubo)

Lymphogranuloma venereum is a chronic, sexually transmitted disease whose main effects result from damage to the lymphatic system draining of the site of infection. The disease is universal but prevalence is much greater in tropical and sub-tropical areas.

Cause
The infectious agent is an intracellular parasite, a member of the *Chlamydia* group. The agent, often described as a virus, differs from viruses by possession of both DNA and RNA. A similar agent is responsible for psittacosis, and the TRIC agent (see p. 206) is another member of the group.

Clinical course
The incubation period is usually between 7 and 15 days, but longer and shorter periods have been reported.

The *primary stage* is a small painless papule or ulcer. It may be found anywhere on the external genitalia but is rarely recognised in women; in men most primary lesions are seen on the shaft or corona of the penis. The lesion is transient and often unnoticed by the patient.

The disease usually presents with uni- or bi-lateral enlargement of inguinal lymph nodes (the *bubo*). The bubo usually occurs after the primary lesion has healed: an incubation period of three to four weeks is most usual. The enlarged glands are matted together and occur both above and below the inguinal ligament (the *groove sign*). Pain in the bubo is usual, but constitutional symptoms are rare. Central softening and multiple fistula formation may ensue.

The late stages of the disease are due to the blockage of lymphatic channels by the infection. Distal oedema develops which may result in gross elephantiasis of the genitals. In females, the vulval elephantiasis is known as *esthiomène*. Ano-rectal stricture formation (and complications) may also occur in late LGV, usually in females but occasionally in homo-sexual males as a result of para-colic lymphatic damage.

256 Electron micrograph of sectioned LGV particles, showing the bacterium-like ultrastructure: including nuclear (N) and cytoplasmic (Cy) regions, and a double envelope representing the cell wall (CW), and plasma membrane (PM).

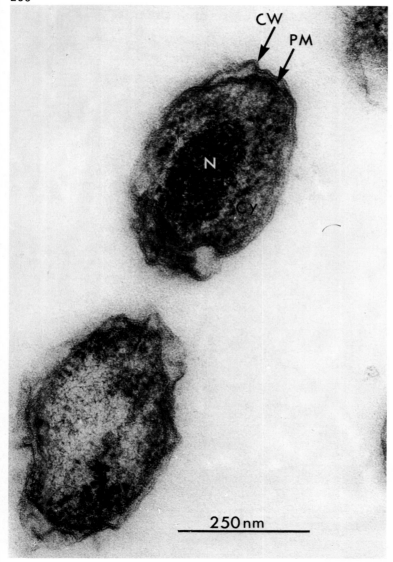

250 nm

Diagnosis

Chlamydiae may be recognised in stained smears of material from lesions or bubo aspirate or by tissue culture.

Serum can be tested for *LGV complement-fixation*; because of cross-reactivity, titres of < 1/40 are not regarded as significant.

The intradermal *Frei test* becomes positive after infection with LGV; positivity is life-long and results must be interpreted with caution.

257 HeLa cell culture infected twenty hours with JH strain of LGV agent. An early cytoplasmic 'inclusion' is present, consisting of a microcolony of initial body particles. Giemsa stain.
258 48 hours stage of the LGV growth cycle. A fully-developed 'inclusion' now almost filled the HeLa cell cytoplasm. It consists mainly of small pink-staining particles which are the infective form of the LGV agent.
259 Frei test: positive result on right forearm, control on left forearm.
260 Primary lesion in coronal sulcus. Note similarity to primary syphilis.
261 Same patient as fig. 260. The discharging bubo is never seen in syphilis.

257

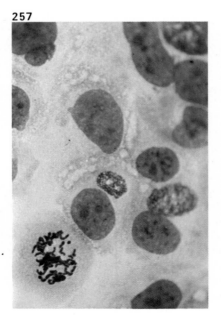

258

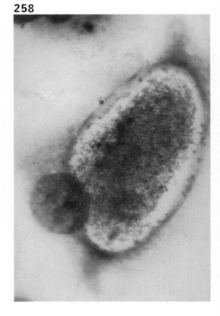

259

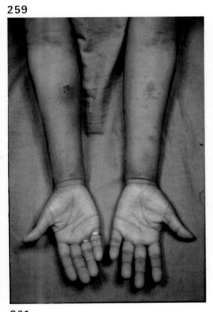

260

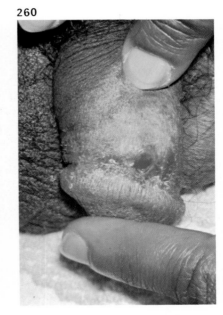

261

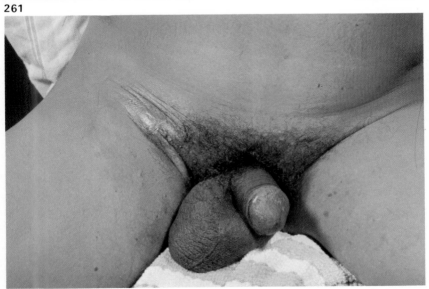

CHANCROID, GRANULOMA INGUINALE
& LYMPHOGRANULOMA VENEREUM

262 Primary ulcer and groin bubo.
263 Bubo of groin. The marker on the penis shows the site of the primary lesion which was almost healed at the time of photography.
264 'Groove' sign of inguinal bubo.
265 Late LGV: elephantiasis of scrotum.
266 LGV of vulva. Note considerable oedema.

262

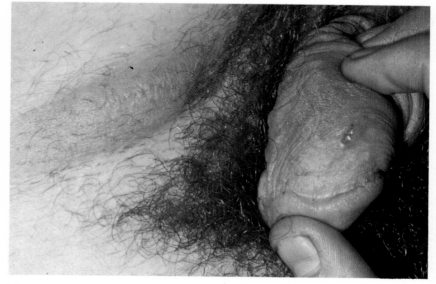

263

264

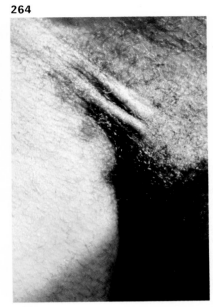

265

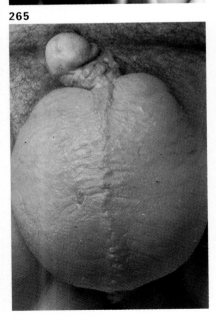

266

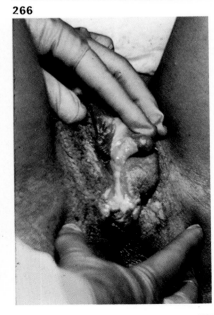

267 Early esthiomène: inner surfaces of labia.
268 Early esthiomène: outer surfaces of labia.

267

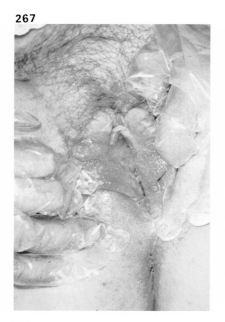

268

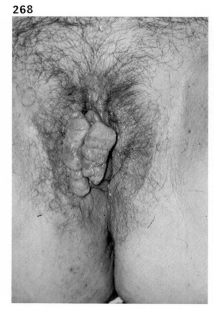

COMMON INFLAMMATORY CONDITIONS

GONORRHOEA

This universally common, highly contagious disease is a bacterial infection affecting columnar or transitional epithelium. It may, therefore, be found in the *urethra* or *rectum* in males; the *urethra, cervical canal* and *rectum* in females; the *pharynx and tonsils* in both sexes; the *conjunctival sac* (especially in the newly born) in both sexes; and in the *vagina* and *vulva* of pre-pubertal females. Local or metastatic spread occasionally occurs.

In most male cases the disease is acute and complication due to local spread of the organism is rare. In females the disease is frequently asymptomatic; in consequence diagnosis and treatment are often delayed and local complications may be the factor which causes the patient to attend.

In both sexes bacteraemia may occur and result in metastatic involvement of remote tissues, *e.g.* skin lesions, septic arthritis or, very rarely, endocarditis or meningitis. In males post-gonococcal stricture of the urethra is now very rare.

The organism
The causative organism is the coccus *Neisseria gonorrhoeae*. It is usually identified in Gram stained smears as an intracellular Gram-negative diplococcus. Positive distinction from other *Neisseria* is made by the unique pattern of sugar fermentation reactions of organisms grown in culture.

Incubation period
The incubation period of gonorrhoea is usually four to seven days. Occasionally, the incubation period may be as little as 24 hours; prolonged incubation of one month or more appears to be coming more common. In male patients, acute symptomatic disease makes determination of incubation easy in most cases, but in female patients and other asymptomatic infections this determination is much less easy.

Diagnosis
Gram stained smears of material from affected sites are examined microscopically. In males most cases can be diagnosed by this method but cultures should always be inoculated to find infections unrecognised on smears. In female patients smear diagnosis is less accurate and a greater proportion of infections can only be demonstrated by cultural methods.

A recent refinement of diagnosis is the use of *fluorescent staining* methods to identify gonococci. Great diagnostic accuracy is achieved but the investigation takes considerable time and at present is unsuitable for routine clinic use.

The gonococcal complement-fixation test (*G.C.F.T.*) is seldom helpful. Reports show that the test currently in use gives unacceptably high rates of false-negative and false-positive results. Considerable research is being undertaken at present to develop a more satisfactory serum test.

269 Gram-stained smear showing Gram-negative intracellular diplococci: *N. gonorrhoeae.*
270 Urethral smear: fluorescent staining; intracellular diplococci.
271 Fluorescent staining of smear prepared from culture showing gonococci and other organisms.

269

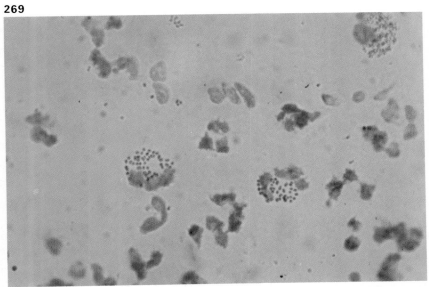

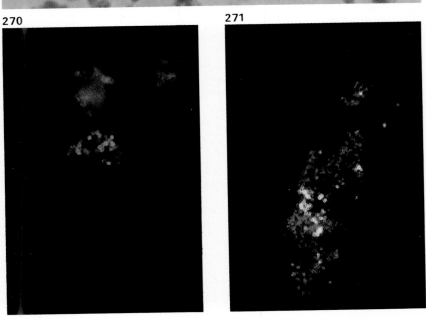

Uncomplicated gonorrhoea in males

The majority of infections are *acute symptomatic urethritis*. Asymptomatic urethritis can occur, and patients may be found to have easily detected signs of infection or the infection may only be detectable on careful examination. *Gonococcal proctitis* is found in passive homosexuals, some of whom are found to have concurrent urethritis.

Clinical course

Gonococcal urethritis. In most cases symptoms progress rapidly after onset. *Pain on urination* is frequent and may become extremely severe and cause retention of urine. *Urethral discharge* begins as scanty mucoid secretion which, in most cases, soon becomes copious and grossly purulent, and is sometimes bloodstained. Moderate *meatal oedema* and *balano-posthitis* are common, and painful *lymph gland enlargement* occurs in about 15 per cent of cases. Slight to moderate oedema of the distal penile shaft and prepuce is frequently found and may be exacerbated by thrombosis of penile dorsal veins or lymphangitis. Rarely, gonococcal exudate may cause folliculitis or cellulitis on the thigh or abdomen. Anatomical anomalies such as para-urethral ducts or median raphe sinuses may sometimes be the site of gonococcal infection. The relative avascularity of these structures may make cure more difficult to achieve.

Gonococcal proctitis. Symptoms are often mild or absent: many patients found to have infection attend at the instigation of a partner who has developed gonococcal urethritis. Patients may complain of *anal dampness, pruritus* or *discomfort,* a mucoid or purulent *anal discharge* and, occasionally, *tenesmus.* White, yellow or bloodstained discharge may be noticed on the surface of the motions.

Proctoscopy may show normal mucosa or a variable degree of *proctitis*; inspection may show no discharge (even in cases later shown by smear or culture to be infected) or muco-pus may be seen. In severe cases, bloodstained purulent discharge is seen to be exuding from mucosal folds.

272 Typical gonococcal urethritis. Note associated meatitis.

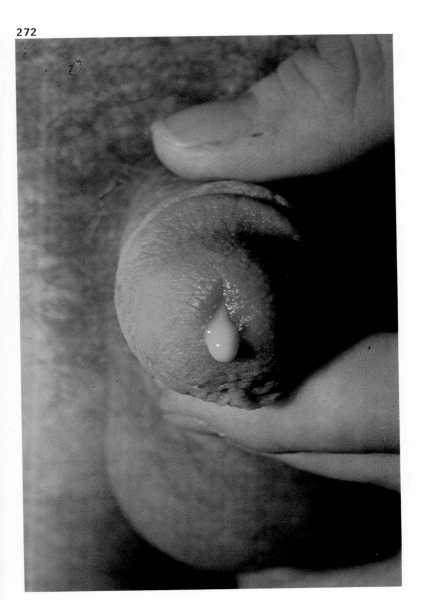

273 Gonococcal urethritis with preputial oedema.
274 Gonorrhoea of median raphe sinus.
275 Gonococcal discharge in hypospadias.
276 Gonococcal urethritis with balanitis.
277 Gonococcal urethritis with oedema of distal penile shaft.
278 Gonococcal urethritis with folliculitis of thigh.

273

274

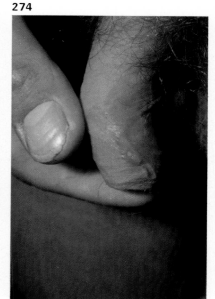

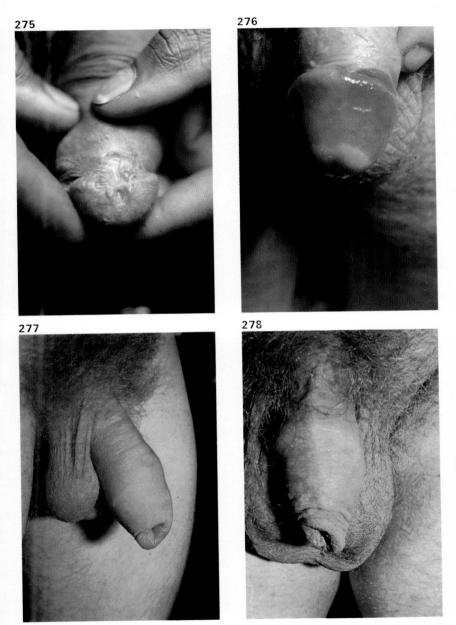

275

276

277

278

279 Gonococcal urethritis complicated by perineal sinus.
280 Gonococcal proctitis: muco-purulent secretion.
281 Gonococcal proctitis: note oedema of mucosa.
282 Gonococcal proctitis: erythema of mucosa and profuse purulent exudate.

279

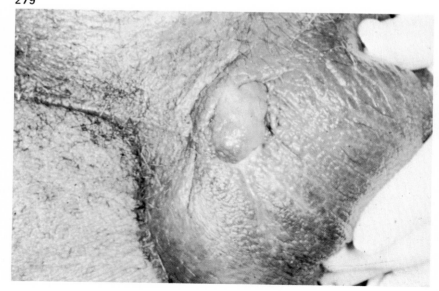

280

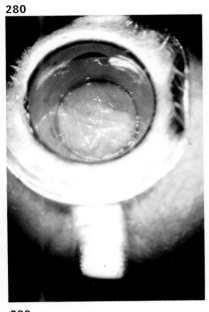

281

282

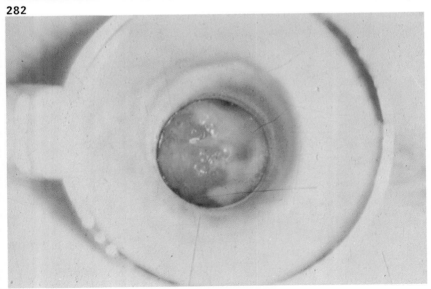

Genital complications of gonorrhoea in males

Spread of gonococcal infection from the urethra to contiguous structures may occur in early infection but such complications are now extremely rare (< 0·5 per cent). Involvement of local glandular structures (*Tyson's, Littre's, Cowper's* and *prostate*) or uni- or bi-lateral *epididymitis* is occasionally seen. Rarely, gonococcal penile ulcers are found. Peri-urethral abscesses and fistulae are now almost unknown.

Urethral strictures developing long after gonococcal urethritis are still occasionally found but are now very uncommon.

Local complications nearly always develop shortly after the appearance of the related urethritis; determination of the aetiology is usually easy. Occasionally signs of urethritis may be slight and the infection difficult to diagnose. It has been observed that when the signs of epididymitis appear the signs of urethritis often diminish. Rarely, the complication may be the only gonococcal lesion present.

283 Gonococcal urethritis with Tysonitis.
284 Gonococcal epididymitis.

283

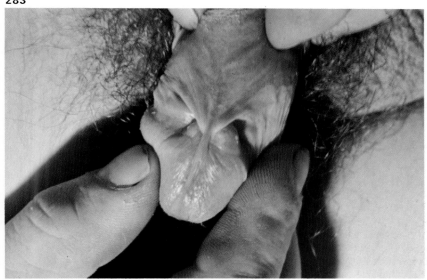

284

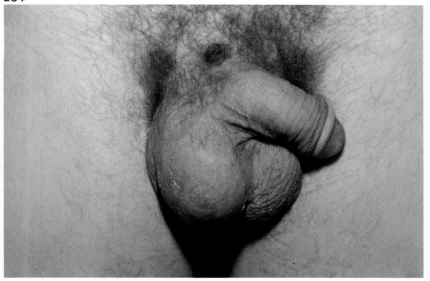

Uncomplicated gonorrhoea in females

Gonococcal infections in females may involve the *urethra, cervical canal* or
rectum; any or all of the sites may be affected concurrently. Gonococcal
proctitis in females may occur as a result of rectal intercourse or by auto-
infection from infected vulval discharge. Proctitis may be resistant to
treatment and may itself cause re-infection of the urethra or cervix.

Clinical course

Symptoms are *absent* or *so slight* that they are ignored by the patient in
about 80 per cent of cases, even in those patients who have recognisable
signs of infection. Symptomatic patients most frequently complain of *urinary
symptoms*—mild to severe pain on urination, dysuria and sometimes
symptoms of cystitis. Slight or moderate increase in *vaginal secretion*
may be noticed; the secretion may become discoloured, often yellow.
Rectal symptoms occur as in male gonococcal proctitis. Rarely, there is
painful enlargement of inguinal lymph nodes.

Signs of infection are very variable. Many patients with bacteriologically
proved infection have a completely normal appearance when examined.
Muco-purulent or purulent *urethral discharge* may be seen (even in the
absence of urinary symptoms), and is often associated with oedema of the
meatus. Ulceration of the meatal margin and involvement of Skene's glands
is occasionally found. The cervix may appear normal with mucoid secretion
or may show endocervicitis, varying from slight erythema around the os to
extensive (2–3 cm diameter) granulating and purulent erosions. *Cervical
discharge* is typically muco-purulent: in severe cases it may be profuse,
frankly purulent and blood-stained. Signs of proctitis are as variable as in
males with gonococcal proctitis.

285 Gonorrhoea: periurethral oedema.
286 Gonorrhoea: purulent 'vaginal' discharge.
287 Gonorrhoea: urethritis.

285

286

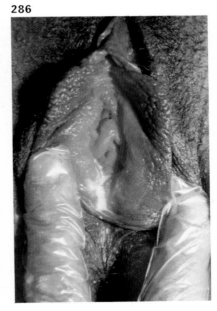

287

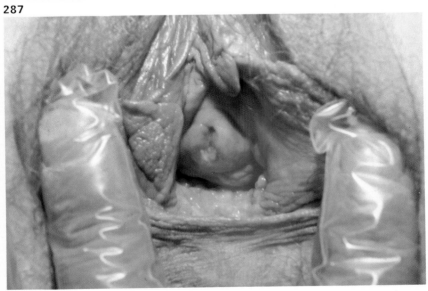

288

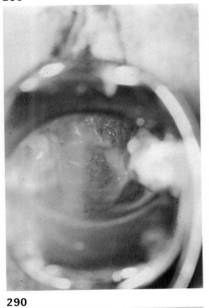

289

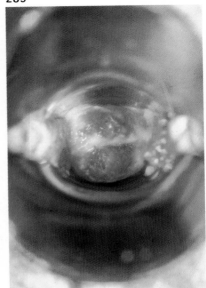

290

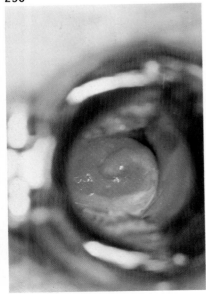

291

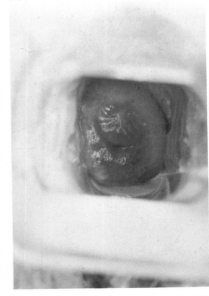

Genital complications of gonorrhoea in females

Complications of gonorrhoea are relatively much more common in females than in males. It has been estimated that the incidence is about 10 per cent, and a considerable number of female patients are found to have gonorrhoea solely because of the development of symptoms or signs of complication. This high incidence is almost certainly due to the asymptomatic nature of most cases of early infection which may allow the infection to be present for a considerable time before manifestation with a consequent greater likelihood of local spread and complication occurring.

As in male patients, local spread may involve local glandular structures (*Skene's* and *Bartholin's*) or progress into or through the uterine cavity to cause *endometritis* or *salpingitis.*

Bartholinitis may vary from slight painless enlargement of the gland, detected incidentally, to a surgical emergency with a grossly enlarged, exquisitely tender gland, with associated cutaneous oedema and erythema. Pus, in which gonococci can be found, may be expressed from the Bartholin's duct or may be obtained by aspiration or drainage of the gland. Following acute infection Bartholin's cyst may occur.

The most serious complication of gonorrhoea is *salpingitis* with possible sequelae of chronic pelvic infection, sterility and increased likelihood of ectopic pregnancy. Gonococcal salpingitis is very variable in severity, ranging from mild cases recognised during examination to severe cases presenting as acute abdominal emergencies. The diagnosis is clinical (sometimes made at laparotomy) ; the aetiology is determined by genital bacteriology. Symptoms and signs include lower abdominal pain (often in one or both iliac fossa) and tenderness ; prolonged and irregular menstruation or intermittent vaginal bleeding ; deep dyspareunia ; and, often, moderate pyrexia and malaise. Bimanual pelvic examination may reveal pain when the cervix is moved (the 'chandelier' sign, see p. 41) or enlargement and tenderness of the Fallopian tubes. In severe cases pelvic peritonitis may be present, and chronic pelvic infection may ensue.

292 Gonorrhoea : Bartholin's abscess.
293 Gonorrhoea : Bartholin's abscess.

292

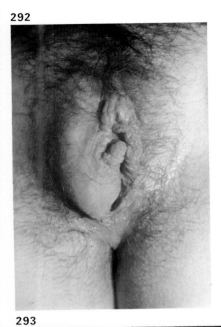

293

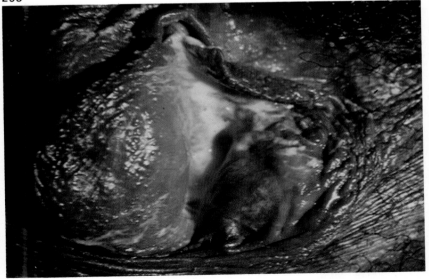

COMMON INFLAMMATORY CONDITIONS

Metastatic gonorrhoea

Neisseria gonorrhoeae, like any other pathogenic organism, can occasionally
find their way into the bloodstream to produce bacteraemia and metastatic
complication. Bacteraemia is considered more likely to occur in chronic
cases or to follow instrumentation or manipulation of infected tissues.
In most reported cases of metastatic complications the genital focus has been
present for two to four weeks at least and may be asymptomatic. Reports
in the past have probably incorrectly ascribed a gonococcal aetiology to
patients with symptoms and findings of Reiter's disease, a complication
of non-specific urethritis.
Metastatic complications are very rare at the present time and are only
mentioned briefly: *septic arthritis, tenosynovitis* (diagnosis of these two is
only tenable if gonococci are demonstrated), *endocarditis, meningitis,
skin lesions.* Skin lesions are the commonest complication, usually seen
on the extremities but occasionally on the trunk. The lesions are transient;
they begin as erythematous papules and progress to vesicles or pustules
which heal without scarring after a few days.

Other sites of gonococcal infection—pharynx and tonsils
Recent reports show an increasing frequency in gonorrhoea affecting the
pharynx and tonsils, apparently linked with an increase in oro-genital
contact (cunnilingus, fellatio). The infection maybe communicated by oral
contacts (kissing) alone. Genito-urinary gonorrhoea can usually be found
as well; in some cases the disease is confined to the oral cavity. The infection
is often asymptomatic and is only discovered by examination. In some
patients symptoms of pharyngitis or tonsillitis occur: examination may show
diffuse pharyngitis, follicular tonsillitis or, occasionally, ulceration of the
throat or tonsil. Moderate cervical lymph gland enlargement is common.
The diagnosis must be established by careful bacteriological examination
including sugar fermentation reactions or by fluorescent methods as other
organisms frequently found in the oral cavity may resemble gonococci on
smears.

294, 295 & 296 Gonorrhoea: skin lesions. Vesico-pustules of foot, finger
and hallux. Note erythematous halo.

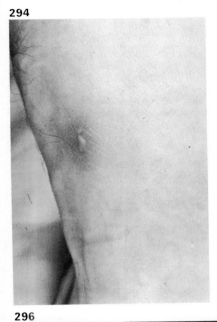

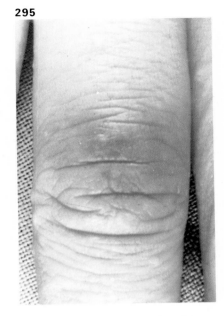

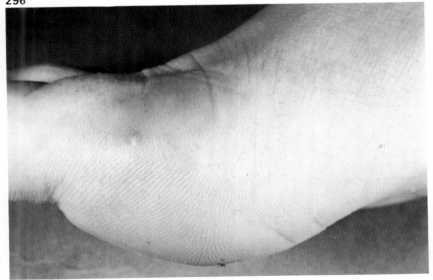

Other sites of gonococcal infection—conjunctival sac

Infants Gonorrhoea is one of the causes of *ophthalmia neonatorum*: it is now rare. Signs of the infection usually appear 2–5 days after delivery, affecting one or both eyes. The infant becomes photophobic and mucoid discharge appears at the lid margins, rapidly becoming profuse, sometimes blood-stained. The eyelids become swollen; erythematous and subcutaneous haemorrhages may appear. The tarsal conjunctiva shows oedema and intense injection; the bulbar conjunctiva shows injection and sub conjunctival haemorrhages are common. In advanced cases corneal ulceration, corneal perforation and panophthalmitis may occur. The infection is acquired during delivery from unrecognised maternal gonococcal endocervicitis; it is essential for the mother and her consort to be examined.

Adults The course of the infection is essentially the same as in infants. The disease is usually acquired by accidental inoculation (often auto-inoculation) of infectious material from genito-urinary gonorrhoea.

Vulvo-vaginitis
Prior to puberty the vulval and vaginal epithelium is susceptible to gonococcal infection. After puberty the hormonal changes causing cornification of the epithelium diminish the susceptibility. Contact with *N. gonorrhoeae* by young females may cause vulvo-vaginitis. In infants the infection may be acquired during delivery, but carry-over of maternal oestrogens usually causes the epithelium to be temporarily resistant. In older girls infection may be acquired by sexual contact (sometimes unadmitted or unrealized); or accidentally from fomites (*e.g.* rectal thermometers); or from non-sexual contact with infected adults. The infection may present with vulval soreness, painful urination, or discharge on underclothing. Examination shows purulent vaginal discharge and vulvitis, and often considerable oedema. Careful bacteriological tests are essential to distinguish gonococcal infection from other causes of vulvo-vaginitis. When gonococcal vulvo-vaginitis is found it is important to trace the source, who may be asymptomatic and unaware of the disease.

297 Gonococcal ophthalmia: adult, with sub-conjunctival haemorrhages.
298 Gonococcal ophthalmia: infant, showing marked oedema of lids.

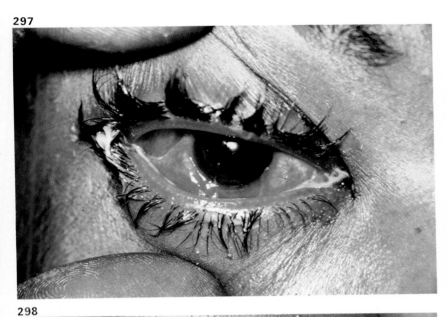

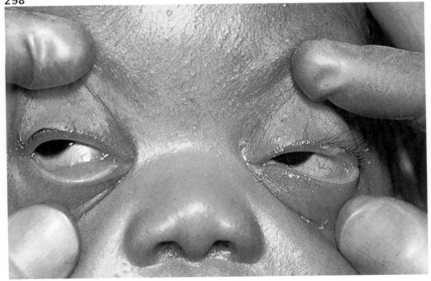

NON-GONOCOCCAL GENITAL INFECTION

Urethritis in male patients (and corresponding conditions in female patients) can be simply divided into two categories: *gonococcal* and *non-gonococcal*. The former has already been dealt with in the preceding section. *Non-gonococcal urethritis (N.G.U.)* can be further sub-divided into groups with known aetiology and unknown aetiology. About 10 per cent of cases of N.G.U. are found to have a recognisable cause (see opposite); these are discussed in separate sections. In the remaining 90 per cent of cases no cause can be identified. The condition is known as *non-specific urethritis (N.S.U.)*, with the female equivalent known as *non-specific cervicitis (N.S.C.)*. *Non-specific proctitis* also occurs. The epidemiology of non-specific genital infections is very suggestive of sexual transmission and although various organisms (*e.g.* mycoplasma, PPLO) have been suggested as aetiological agents no definite links have been established.

Two further conditions have also to be considered in this section: *acute haemorrhagic cystitis* and *chronic prostatitis*.

Acute haemorrhagic cystitis The condition presents as non-gonococcal urethritis associated with haematuria (often gross) and severe symptoms of cystitis. The cause is unknown but the clinical pattern suggests that the condition is a separate entity.

Chronic prostatitis The definition, diagnosis and effects of this condition are all vague as the symptomatology may occur in patients with normal physical findings and abnormalities in the prostatic fluid may be found in apparently normal individuals. Findings designated as chronic prostatitis are frequently found in patients with *Reiter's disease, anterior uveitis* and *ankylosing spondylitis* but the relationship of the prostatic to the other findings is not understood. Chronic prostatitis may be a sequel to urethritis of any aetiology and it seems probable that some cases of apparent non-gonococcal urethritis are in fact exacerbations of chronic prostatitis.

The diagnosis is usually based on examination of the prostato-vesicular secretions expressed by prostatic massage (see p. 64). Prostatic fluid is collected on a slide (or series of slides) and examined microscopically. Abnormal prostatic fluid shows excessive numbers of leucocytes ($>$ 5 per 1/12 in. field) which are often clumped. Very rarely organisms such as *T. vaginalis* and bacteria may be identified. Unfortunately, similar findings are not uncommon in apparently normal individuals and it is usually difficult to correlate clinical symptoms and findings with results of prostatic fluid examination.

The symptoms associated with chronic prostatitis include perineal discomfort, occasional mucoid urethral discharge which often occurs with defecation. Urinary symptoms include those of mild obstruction and irritation with urination. Many patients with such symptoms have apparently normal prostatic findings on examination.

The findings on examination include slight enlargement of the prostate (and often of the seminal vesicles also) traditionally with a 'boggy' or 'nodular' feel, but such findings may occur in asymptomatic men or with normal microscopical examination of the expressed prostatic fluid in cases with symptomatology suggestive of chronic prostatitis.

Causes of urethritis

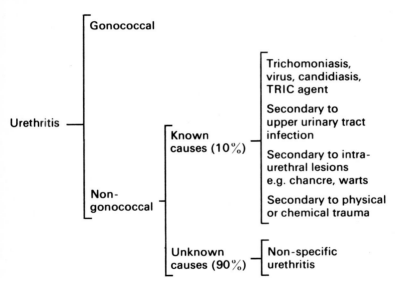

TRIC agent infections

TRIC agent (*TR*achoma *I*nclusion *C*onjunctivitis) is a member of the *Chlamydia* group. Known for a long time to produce a characteristic conjunctivitis in the neonatal period, it has recently been found to be present in the cervical canal of some mothers of affected infants and in the urethra of some of the fathers. Using special techniques the TRIC agent can be found in a proportion of patients presenting with non-gonococcal urethritis, cervicitis and proctitis. The exact role of the TRIC agent in genital infection has not yet been defined, but the evidence suggests that it may be the cause of infection in up to 40 per cent of cases. Clinically, follicular lesions of similar morphology have been observed in the urethra, cervix, rectum and conjunctivae but at present clinical examination findings are suggestive but not pathognomonic. The organism is not identified by current methods of routine bacteriological examination.

299 Microphotograph of TRIC agent: Halberstaedter–Prowazek inclusion.
300 'Initial body' inclusions of TRIC agent in epithelial cells.
301 Cervical canal showing follicular lesions.
302 Urethral meatus showing meatitis, papillary congestion and follicular lesions.

299

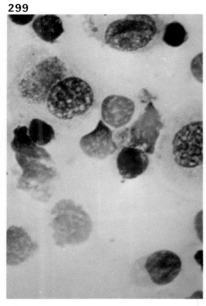

300

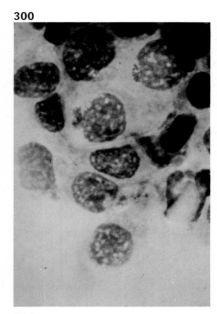

301

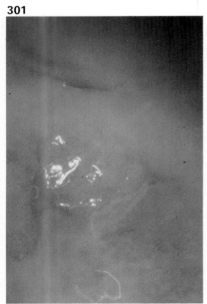

302

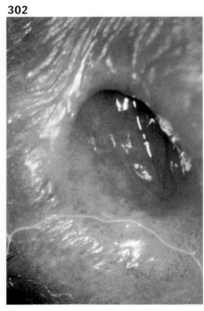

NON-SPECIFIC GENITAL INFECTION (N.S.U.)

Non-specific genital infection is already extremely common and incidence appears to be increasing rapidly; in England, N.S.U. is now more frequent than gonorrhoea. Most cases of non-specific genital infection are mild but a small proportion develop the potentially serious complication, *Reiter's disease*. The infectiousness of the condition is unknown. Symptoms and abnormal findings occur almost exclusively in male patients but their apparently normal female partners appear to be carriers of the infectious agent as, if they are left untreated, the male is likely to show signs of recurrence of infection when intercourse is resumed. Male passive homosexuals may also be asymptomatic carriers.

Clinical course

In *males* the incubation period is usually two to three weeks. Some infections present within a few days of contact and in mild cases the symptoms may not be noticed for several months.

Symptoms are usually mild but are occasionally severe. Some cases are completely asymptomatic and are only detected on routine examination; occasionally the development of a complication is the first sign of infection. *Urinary symptoms* vary from mild meatal irritation to severe dysuria with or without urgency and frequency of urination. *Urethral discharge* is usually noticed by the patient but in mild cases may only be detected after the urine has been retained overnight. The discharge is usually mucoid or mucopurulent but occasionally may simulate typical gonococcal urethritis. Associated *balano-posthitis* and local *lymphadenitis* are uncommon. Rarely, local complications such as *epididymitis* may occur; urethral stricture occasionally develops later. Following resolution of urethritis prostatic fluid often shows excessive numbers of leucocytes but it is uncertain whether this finding is primary or secondary.

In passive homosexual patients *non-specific proctitis* may be found. This is usually asymptomatic but may cause rectal discharge or tenesmus.

In *females* complete absence of symptoms and recognisable abnormal findings on examination is usual. Occasionally the development of a complication such as Bartholinitis or salpingitis indicates the presence of non-specific genital infection.

303 Non-specific urethritis: typical muco-purulent discharge.
304 Non-specific epididymitis with inflammatory hydrocele and erythema of scrotum.
305 Non-specific cervicitis.
306 Non-specific cervicitis.

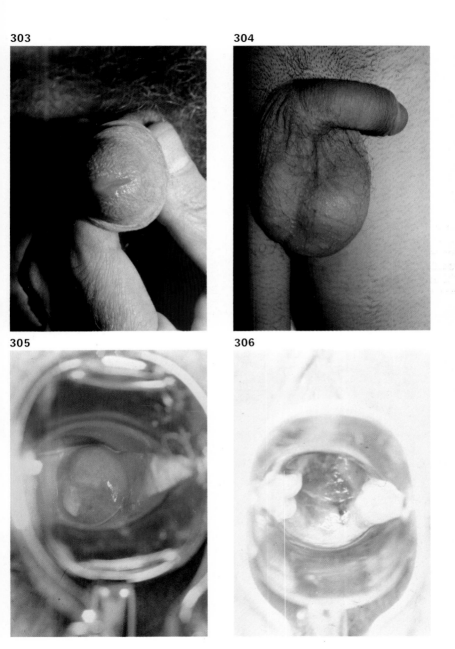

Diagnosis

The diagnosis is made by exclusion of known causes of the presenting symptoms. Gram-stained urethral smears taken from males with N.S.U. show leucocytes and epithelial cells, and may be either abacterial or contain normal urethral commensal organisms; the two glass test shows anterior urethral involvement. Urethral cultures are either sterile or grow commensal organisms only. In female patients leucocytes are often seen in urethral and cervical smears but diagnostic criteria have not been established to indicate whether these findings have significance.

Reiter's disease (Feissinger–Leroy syndrome)

Classically, *Reiter's disease* is a triad of *non-specific genital infection, arthritis* and *conjunctivitis*. The skin and mucous membranes are frequently involved. The classical syndrome is comparatively rare and it is probable that many cases with incomplete forms of the syndrome present in non-venereological departments and are unrecognised. Reiter's disease has a marked tendency to relapse after the initial attack has cleared up. Relapses may feature one or more of the manifestations and are not necessarily related to further episodes of genital infection.

Reiter's disease develops in one to two per cent of patients with non-gonococcal genital infection. It is not clear whether it is due to individual susceptibility; there is some evidence of familial propensity which favours the latter theory. The disease may occasionally develop after dysentery. For unknown reasons this form of the disease appears to be uncommon in the United Kingdom. Males are found to have the disease much more frequently than females (ratio 10:1). This may be because the genital infection is usually overt in males and covert in females, who may present, as mentioned above, with non-genital features of the syndrome in other departments.

Diagnosis

The diagnosis of Reiter's disease is clinical and radiological. There is, at present, no laboratory investigation which is diagnostic, but investigations can positively exclude other conditions which can be confused with Reiter's disease.

Clinical course

In the majority of cases the first indications of Reiter's disease are the appearance of *conjunctivitis* and/or *arthritis* one to two weeks after the onset of genital infection. Unfortunately, the preceding genital infection has no particular features which enable Reiter's disease to be identified in advance, and the severity of the genital infection bears no relationship to the severity of subsequent Reiter's disease. The clinical course is very variable and although most cases are mild, severe systemic disorder is occasionally seen. The disease has protean manifestations which are outlined below. In addition to localised symptoms and signs, systemic abnormalities are often found. In many patients there is moderate fever and the E.S.R. is raised, sometimes to 100 mm or more fall per hour in episodes of disease activity.

The manifestations of Reiter's disease

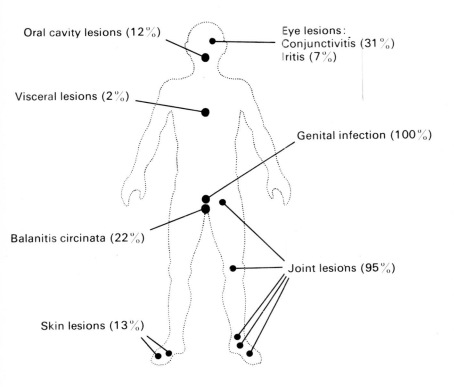

Oral cavity lesions (12%)

Eye lesions:
Conjunctivitis (31%)
Iritis (7%)

Visceral lesions (2%)

Genital infection (100%)

Balanitis circinata (22%)

Joint lesions (95%)

Skin lesions (13%)

Genital infection
The non-specific genital infection found may present as *urethritis, cystitis* or *prostatitis.* Sometimes the infection is asymptomatic and only detected on careful examination. Occasionally extra-genital manifestations of Reiter's disease may precede signs of the genital infection.

Visceral lesions
Thrombo-phlebitis, peripheral neuritis, myocarditis, aortic incompetence and secondary amyloidosis have occasionally been described in patients with Reiter's disease : all are rare. Very rarely, other types of systemic involvement have been reported.

Joint lesions
About 95 per cent of patients with Reiter's disease develop lesions in the joints or adjacent tissues. Usually objective changes of *arthritis* are found but in some patients *arthralgia* is the only abnormal skeletal manifestation. *Arthritis* usually appears in the course of the first episode but may only become evident in a later relapse. With repeated relapses considerable permanent change may occur, often involving the metatarso-phalangeal joints or causing spur formation beneath or behind the os calcis. Occasionally spinal changes, difficult to differentiate from ankylosing spondylitis, may develop. The arthritis is usually poly-articular, often more or less symmetrical and particularly likely to affect the peripheral joints of the lower limbs; the sacro-iliac, knee, ankle and foot joints are most commonly involved. Mono-articular arthritis may occur and is seen most often in the knee. The arthritis is non-suppurative and may be either acute or sub-acute. When it is acute the process may spread to involve adjacent soft tissues, seen most often in the plantar fascia or Achilles tendon. In the majority of attacks the arthritis clears up after four to six weeks. Relapses are usually shorter in duration and milder but with successive attacks increasing deformity is likely to develop.

307 Reiter's disease: arthritis of left hand, principally affecting metacarpo-phalangeal joints.
308 Reiter's disease: arthritis of foot. Note involvement of mid-foot, ankle and metatarsals.
309 Radiograph of sacro-iliac joints. Note sclerosis on left side.

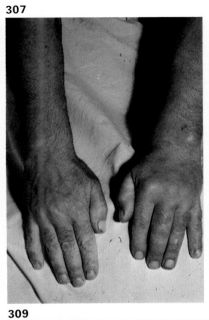

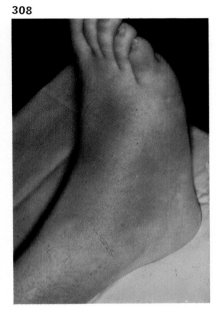

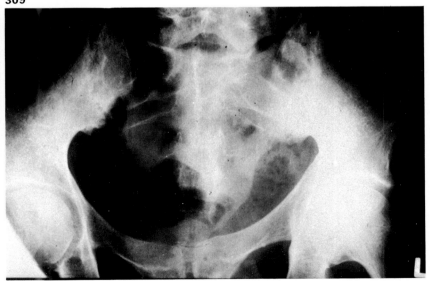

Eye lesions

Conjunctivitis occurs in about 30 per cent of patients with Reiter's disease and is often the first indication of the condition. It is usually mild and transient but may occasionally be severe and associated with chemosis. One or both eyes may be involved. The tarsal conjunctiva tends to be more severely affected, especially at the lateral angles. Relapses are common, and conjunctivitis is frequently the only lesion in a relapse.

Iritis occurs in about eight per cent of patients. It may initially be found late in first attacks of Reiter's disease or may only appear weeks or months after initial symptoms; relapses are very common. Synechiae formed in repeated attacks can cause secondary glaucoma which may endanger vision; iridectomy is sometimes necessary to preserve sight.

310 Radiograph of foot. Note inferior calcaneal spur, posterior calcaneal erosions and deformity of metatarso-phalangeal joints.
311 Reiter's disease: bilateral conjunctivitis, affecting bulbar and tarsal conjunctivae. Note more marked involvement of the lateral angles.
312 Severe bilateral conjunctivitis with chemosis. Such severity is unusual.

310

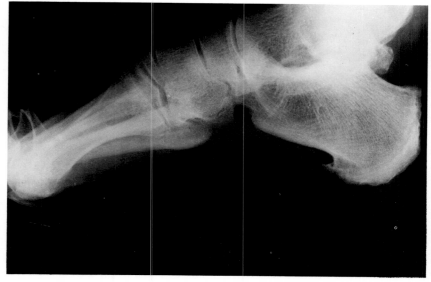

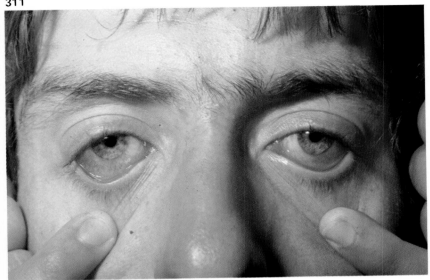

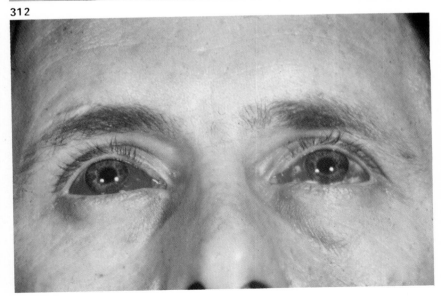

Skin lesions

Keratodermia blennorrhagica occurs in about 15 per cent of patients, often in the more severe cases with many signs, but is occasionally seen alone. The rash is found most characteristically on the soles of the feet. It may be found elsewhere on the skin or scalp, and in particularly severe cases the nails and nail beds are likely to be involved. Typical lesions occur on the penis in circumcised males. The lesions begin as discrete small subcutaneous papules which become vesicular and covered with a heaped-up adherent crust. Spreading lesions become confluent and may involve extensive areas. Sub-ungual lesions may cause lifting or loss of the nail.

Balanitis circinata is seen on the glans penis and prepuce in uncircumcised males. Small, discrete, round or oval red macules or shallow erosions appear which often spread centrifugally and become confluent. The topography of the confluent lesions is circinate and the margin is often slightly elevated and grey in colour. About 12 per cent of patients with Reiter's disease have this lesion.

313 Keratodermia blennorrhagica of forehead and scalp.
314 Keratodermia blennorrhagica of legs with effusions in knees.
315 Severe keratodermia blennorrhagica of feet. Note 'pustular' lesions and hyperkeratosis.
316 Severe keratodermia blennorrhagica of feet: exfoliating.

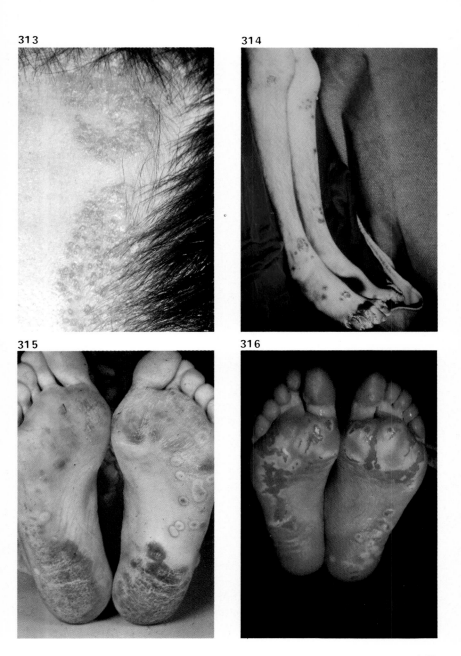

313

314

315

316

317 Balanitis circinata of penis: mild lesions. Note similarity to fungal infections.

318 Typical balanitis circinata. Note topography.

319 Balanitis circinata: the lesions are larger than usual.

320 Circumcised penis showing keratodermia blennorrhagica.

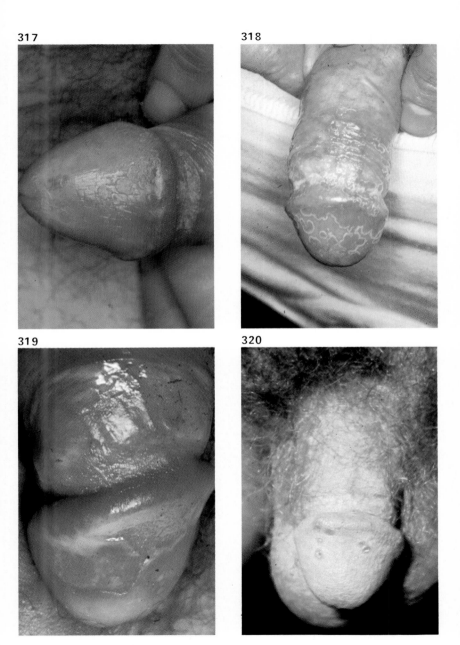

Oral lesions

Small painless vesicular or erosive lesions are found in the oral cavity in about 12 per cent of patients with Reiter's disease. Lesions are commonly seen on the tongue (dorsum and edge) or palate but may also occur elsewhere in the oral cavity. The surface of the lesion may be erythematous or covered with grey exudate. This exudate is often thicker at the margin of the lesion.

321 Vesicular lesions and erythema of hard palate and lesions on cheek.
322 Extensive superficial erosion of tongue.
323 Patchy erosion of tongue.

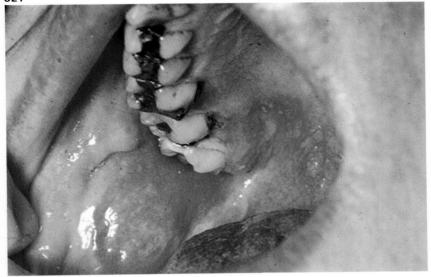

321

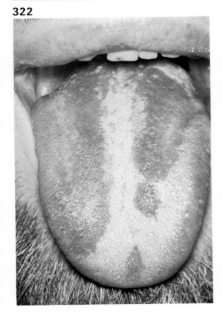

322

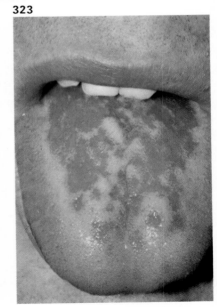

323

CANDIDIASIS (THRUSH)

Candida albicans (monilia) is the most common fungus infection encountered in venereological practice. Genital dermatomycosis may also be
due to other *candida* species (which are not discussed separately) or to the
other fungal infections described on page 258. Thrush may be transmitted
or exacerbated by sexual contact but most infections (particularly in
women) are due to auto-inoculation from the rectum. Symptoms may be
due to hypersensitivity or infection. If due to hypersensitivity alone,
bacteriological investigations are negative. The fungus is frequently found
as an asymptomatic saprophyte in the mouth, rectum and genital regions.
Symptomatic infections are much less common and may be due to the
intervention of another factor. These other factors include pregnancy,
glycosuria, antibiotic or immunosuppressive therapy, and oral contra-
ceptives : physical factors such as obesity and hyperhydrosis may be
significant. In clinical practice the condition may present with symptoms
in one partner only but it is essential to examine the asymptomatic
partner to reduce the chance of re-infection.

The organism
Candida albicans may be found in either mycelial (hyphae) and/or spore
forms. Mixed forms are common; the spores are often seen to be
budding. In Gram-stained smears taken from appropriate sites *C. albicans*
is nearly always Gram-positive. The fungus may also be identified on culture.

Diagnosis
The diagnosis is made by demonstration of the causative organism,
C. albicans, either on smear or more frequently on culture. The organism
may also be recognised on cervical cytological smears. The material for
examination may be taken from any of the sites liable to be affected. It should
be remembered that symptoms of candidiasis may be due to contact
dermatitis; these patients will be bacteriologically negative. It is *essential*
to examine for glycosuria as candidiasis is a not infrequent presentation of
diabetes.

Clinical course
Asymptomatic or symptomatic infections may be found; both are more
frequent in women. The incubation period is impossible to determine as
asymptomatic infestation is so common. Sometimes symptoms may appear

324 Hyphae of *Candida albicans* in vaginal smear : Gram stain.
325 Dormant and budding spores in vaginal smear : Gram stain.

324

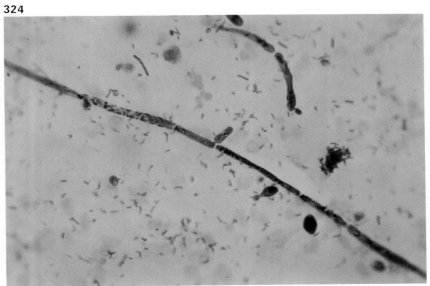

325

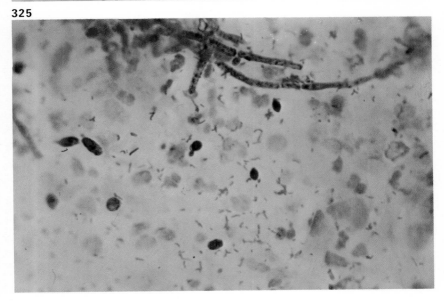

within minutes of sexual contact (due to hypersensitivity) or may be
delayed for several days while infection develops.

Candidiasis in women
The pattern is very variable and symptoms and findings often do not
correlate. *Irritation (pruritus)* almost always occurs in symptomatic cases
which usually affects the *labia* and *vulva*. It may also affect the perineum,
natal cleft and the groins. Lumpy *vaginal discharge,* scanty or copious, is
another very common symptom. *Swelling* of the labia and *superficial*

326 Candidiasis: typical vulvitis and perivulvitis. Note the oedema, the
colour of the lesion and the perineal and perianal involvement.
327 Candidiasis: vulvitis. Note the oedema and dried yellow discharge.
328 Candidiasis: vulvitis and crural intertrigo.
329 Candidiasis: vulvitis. Note the oedema and the grey mucoid discharge.
330 Candidiasis: vulvitis. Note excoriations.
331 Candidiasis: vulvitis. Close-up view showing peripheral pustular
lesions.

326

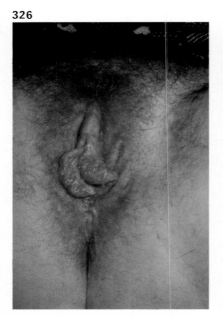

327

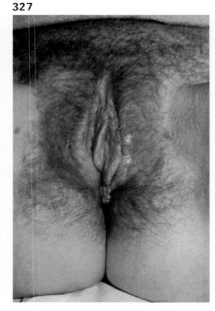

328

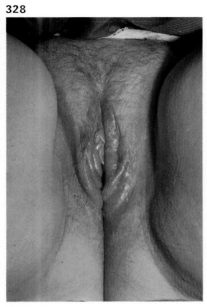

329

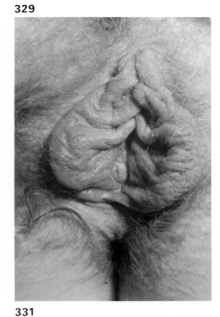

330

331

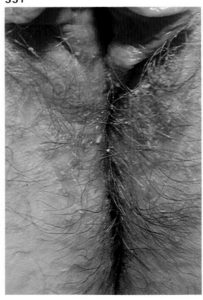

dyspareunia are less frequent symptoms; occasionally a *rash* of the labia or groin is noted. Some women complain of a *'sour' genital odour.*

Findings on examination are equally variable. The labia are often mildly or moderately *oedematous* and may be *erythematous. Intertrigo* often extends to the perianal region and occasionally to the groins. Follicular lesions may be seen, especially at the edges of involved areas. *Vulvitis* may be mild or severe; there is frequently associated *vaginitis* but either may be seen alone. Plaques of typical yellowish cheesy exudate may be seen on the vulva, vagina and cervix. Vaginal discharge may be typical or may be thin and grey. Shallow erosions, usually due to scratching, may be found on the labia and perineum.

332 Candidiasis: chronic infection in Negro patient showing oedema, lichenification and extensive hyperpigmentation of upper thighs.
333 Candidiasis: vulvitis with typical 'cheesy' plaques.
334 Candidiasis: vaginitis with muco-purulent discharge.
335 Candidiasis: vaginitis, with exocervicitis and plaques of typical exudate on vaginal wall.

332

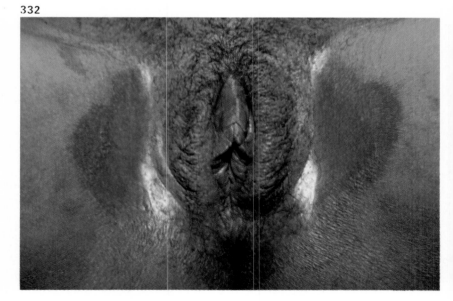

333

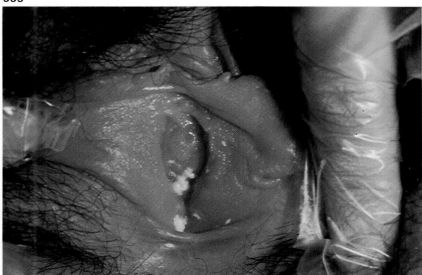

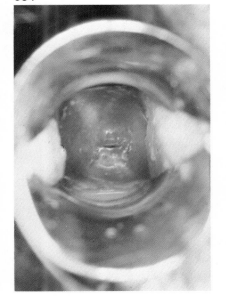**334**

**335**

336 Candidiasis: vaginitis with generalised congestion and muco-purulent discharge. Note the annular lesion on the cervix.
337 Candidiasis: vaginitis with typical plaques of exudate on the cervix and vaginal wall.

336

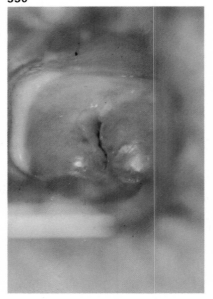

337

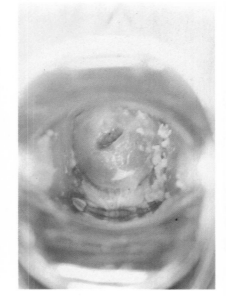

Candidiasis in men

The most common complaint is *irritation* of the *glans* and *prepuce,* often associated with a *rash* and *swelling* of the affected parts. Occasionally the irritating rash may affect the scrotum and groin and, in homosexual patients, the perianal region. Symptoms of urethritis with marked meatal and distal urethral irritation occasionally occur; other symptoms are rare.

Examination shows follicular or diffuse erythema of the glans and prepuce, frequently most marked in the coronal sulcus. Lesions in uncircumcised patients are moist but in circumcised patients they are often dry and show a typical fungal centrifugal pattern. Scaling of the skin or mucous membrane and muco-purulent exudate are often seen; typical plaques are less common. There is usually some oedema present, and in severe cases phimosis may occur. Diffuse erythema of the scrotum and groin, often with marginal folliculitis is occasionally seen. Urethritis may be found and is almost always mild.

338 Candidiasis: balano-posthitis. A dry lesion showing 'glazed' oedematous mucosa and superficial fissuring.
339 Candidiasis: balano-posthitis, showing localised lesion.
340 Candidiasis: balano-posthitis. Typical, showing oedema, exudate and patchy erosions.

338

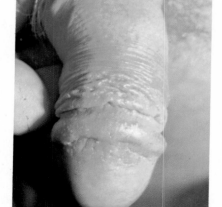

339

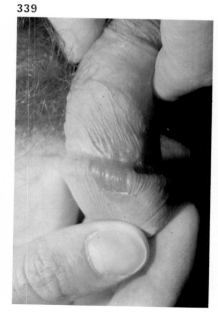

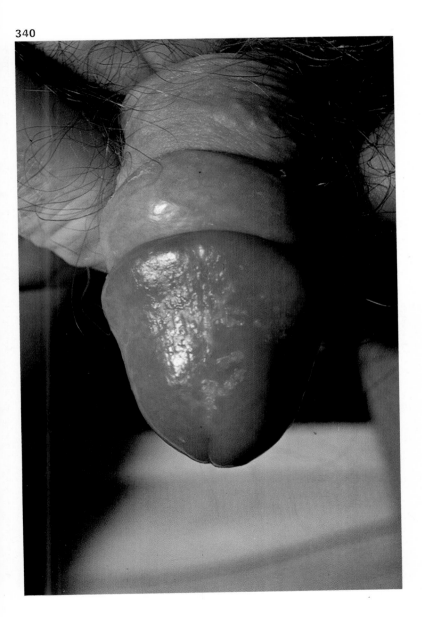

341 Candidiasis: balanitis: circumcised penis with dry patches reminiscent of lichen planus.
342 Candidiasis: balano-posthitis, 'cobblestone' type.
343 Candidiasis: balano-posthitis, adherent perimeatal plaque.
344 Candidiasis: balano-posthitis. Diabetic patient presenting with this lesion. The blood sugar was 390 mg %. Note the typical plaques.
345 Candidiasis: balano-posthitis. Dry lesion showing desquamation and white exudate.
346 Candidiasis: balano-posthitis, diffuse type.

341

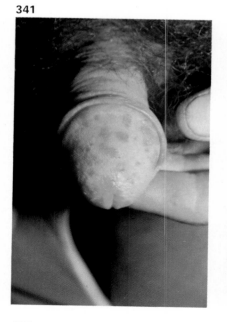

342

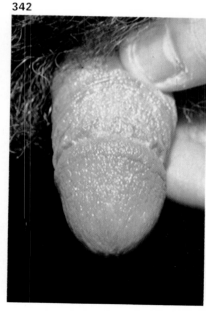

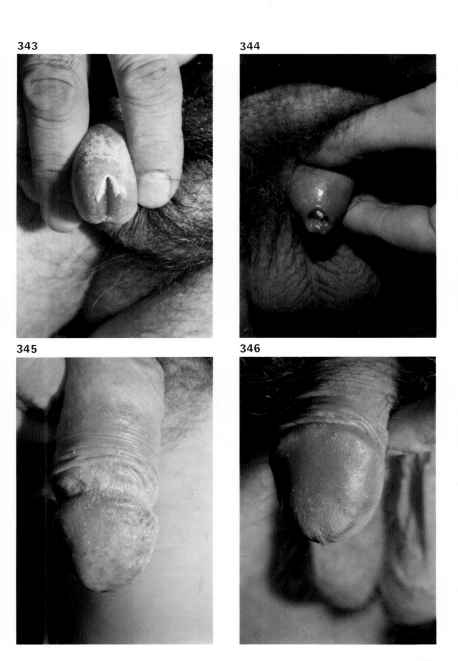

343

344

345

346

347 Candidiasis: diabetic patient, with oedema and phimosis resulting from chronic infection.
348 Candidiasis: associated with herpes and phimosis (Cf. secondary syphilis, fig. 151).
349 Candidiasis: perianal infection in passive homosexual. Note the peripheral satellite lesions. This patient's active partner had candidal balano-posthitis.
350 Candidiasis: satellite pustules at periphery of groin lesion.
351 Candidiasis: glans penis and nails.

347

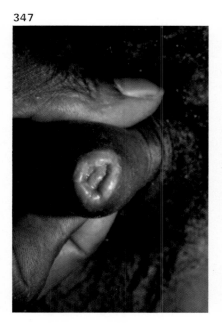

348

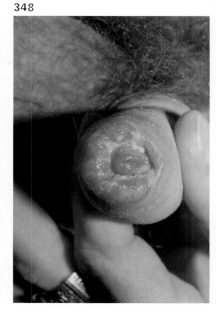

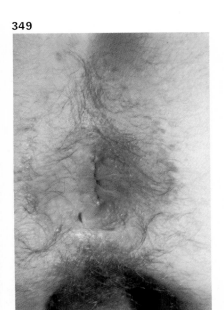

349

350

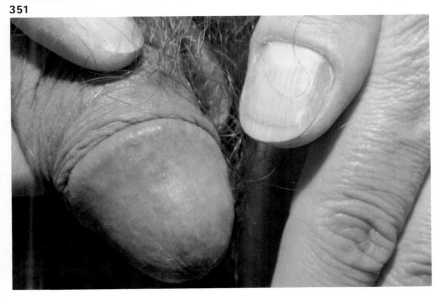

351

352 Candidiasis: intertrigo of scrotum and groin.

352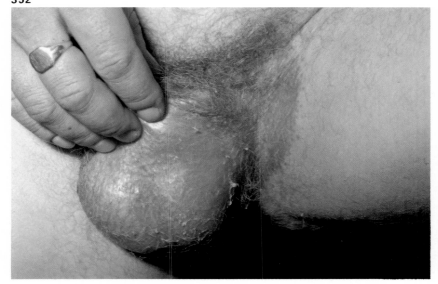

TRICHOMONIASIS

Trichomonas vaginalis is a protozoal parasite which may be found in the *vagina* or *urethra* in women and the *urethra* or *preputial sac* in men; it is occasionally found in other genito-urinary locations. The infestation may be asymptomatic or may cause acute or sub-acute inflammatory changes. It has been estimated that 10 to 15 per cent of women between the ages of 15 and 45 harbour the parasite but the incidence of symptomatic disease is very much less. It is probable that the parasite is usually transmitted sexually but accidental infections can occur. It seems likely that in some cases diagnosed in young female adolescents latent infestation has been present since birth. In women the duration of infestation is usually prolonged in the absence of treatment; in men infestation is often transient. In clinical practice this is reflected by the condition being found 10 times more frequently in women. Trichomoniasis is frequently associated with other genito-urinary disorders and it is essential to undertake examination to exclude such conditions.

The organism
T. vaginalis is a rounded or oval unicellular organism 15 to 30 μ in length. Four motile flagella project at the anterior end and an undulating membrane extends along the side of the body towards the projecting axostyle.

Diagnosis
The parasite is most easily recognised in fresh microscopic specimens. Secretions from the posterior vaginal fornix and/or the urethra in women or from the urethra in men are mixed with a drop of saline on a slide. The specimen is covered with a cover slip or examined by a hanging drop technique. Microscopy, using dark-ground or reduced transmitted illumination, clearly shows the actively beating flagella and the undulating membrane. In affected women the parasite is usually abundant; in affected men it may be sparse, and diagnostic accuracy may be improved by instilling a few drops of saline into the terminal urethra, massaging the area and collecting the expressed fluid for examination.

Staining techniques are unsuitable for routine clinic use, but many unsuspected cases are found when the parasite is recognised in cervical exfoliative cytological smears. Cultures, employing suitable media, inoculated with material taken from the usual sites of infestation, add considerably to the numbers of cases found by microscopy.

Clinical course
In acute and sub-acute cases the incubation period appears to average one to two weeks. In asymptomatic cases this is impossible to determine,

and it is usually impossible to distinguish between latent infestation becoming active and newly acquired infection. Some infestations become evident when another type of infection is superimposed or after trauma to asymptomatic, infested sites. Pregnancy appears to predispose towards infection.

Trichomoniasis in women

Symptoms and findings are extremely variable and not necessarily correlated. Severe symptoms may be associated with almost normal findings and vice versa, the fastidiousness of the individual patient being an important factor. The gradually increasing symptoms in prolonged infestations may be regarded by the patients as 'normal' even when grossly abnormal signs are found on examination. Common symptoms include: increased *vaginal discharge* (sometimes only noticed immediately before or after menstruation); *vulval soreness* and *superficial dyspareunia*; *offensive odour*; *vulval* and/or *perivulval 'rash'* (sometimes involving upper thighs and groin); *vulval* and/or *perivulval irritation*; *vulval* or *labial swelling*; and, occasionally, symptoms of urethritis or cystitis, mild pelvic discomfort and aching in the groins.

Examination findings are equally variable. *Vulvitis, vulval erosions, labial oedema, urethritis* and *vaginitis* are frequently found and *perivulvitis* is common; groin intertrigo is occasionally seen. In severe cases the vaginal walls and cervix show the classical *'strawberry'* appearance with punctate bleeding erosions; in less acute infections there is diffuse erythema of the exocervix. Vaginal discharge is most typically thin, frothy and purulent, but all types of vaginal discharge (mucoid, muco-purulent or grossly purulent and bloodstained) may be observed; the discharge is usually alkaline with pH 7 to 9. Some urethral discharge is often present, and Bartholin's and Skene's glands are occasionally involved.

353 Trichomoniasis: *Trichomonas vaginalis*. Note the pyriform shape, the four flagella, the lateral undulating membrane, the axostyle and the single nucleus.

354 Trichomoniasis: dark-ground preparation showing *T. vaginalis* and epithelial cells.

355 Trichomoniasis: microphotograph of cervical smear (Papanicolaou stain) showing epithelial cells, leucocytes and *T. vaginalis*. The pleomorphic trichomonads are identified by small red spots within the cytoplasm (arrowed).

356 Trichomoniasis: typical vulvitis. Note the classical frothy purulent discharge.

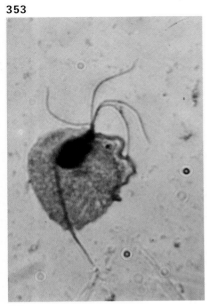

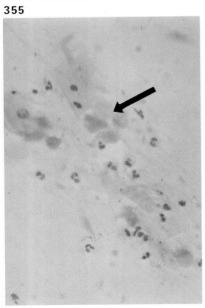

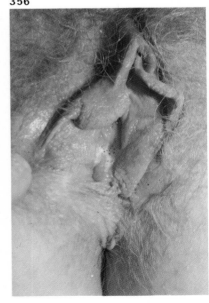

Trichomoniasis in men

The infestation may be asymptomatic and only diagnosed when careful examination is made. Symptomatic cases usually present as *non-gonococcal urethritis*, clinically indistinguishable from N.G.U. due to other causes. Rarely, *T. vaginalis* may cause *balano-posthitis, prostatitis* or *epididymitis*. Symptoms are almost invariably mild: *urethral discharge* (sometimes noticed in the morning only); *urethral irritation;* or symptoms of any complication. Examination of the urethra shows scanty mucoid (often characteristically grey in colour) or muco-purulent discharge in most cases. Other abnormal findings, apart from a 'two glass' test indicating anterior urethritis, are rare. It is important to remember that trichomoniasis in men is frequently

357 Trichomoniasis: vulvitis, severe perivulvitis and perianal intertrigo.
358 Trichomoniasis: perivulvitis with urticarial lesions on the thigh and vulval oedema. Compare with fig. 326 (candidal perivulvitis).
359 Trichomoniasis: acute vulvitis.
360 Trichomoniasis: vulvitis with marked oedema. Compare with fig. 327, candidiasis.

357

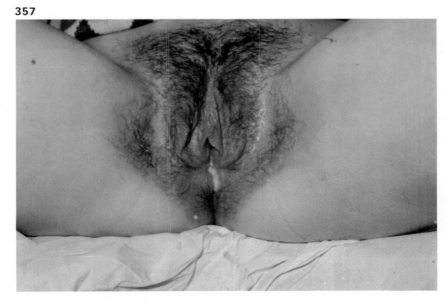

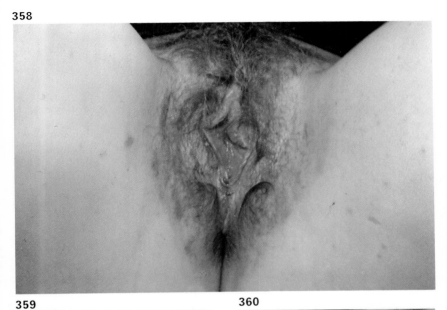

358

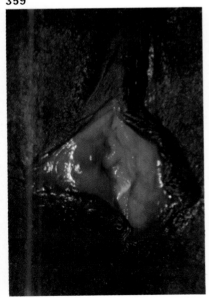

359

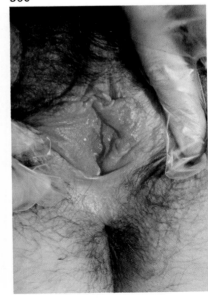

360

associated with structural abnormalities such as urethral stricture and to investigate appropriately whenever the infestation is found.

361 Trichomoniasis: vulvitis. Note oedema and erythema of labia minora.
362 Trichomoniasis: vulvitis, demonstrating frothy exudate of vulva, urethra and introitus.
363 Trichomoniasis: vaginitis and exocervicitis with classical frothy discharge.
364 Trichomoniasis: vaginitis with thick purulent discharge in posterior fornix.

361

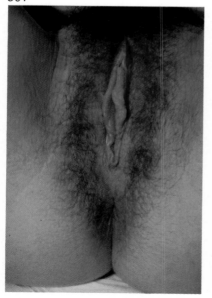

362

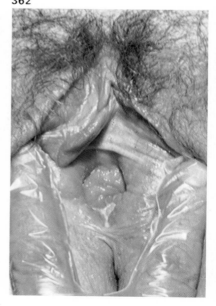

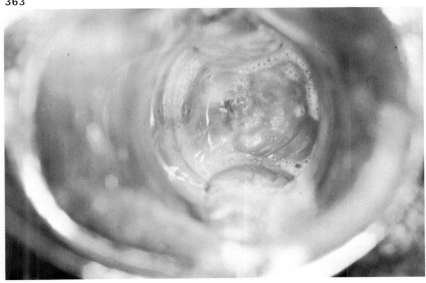

BALANO-POSTHITIS

Balanitis (inflammation of the glans penis) and *posthitis* (inflammation of the prepuce) frequently occur concurrently; balanitis alone may occur in the circumcised but is relatively uncommon. *Balano-posthitis* is seen very frequently, and is often secondary to other conditions. The causes of balano-posthitis are outlined opposite; many of the conditions listed are discussed separately. Non-specific balano-posthitis is usually mild but can occasionally be severe. Many cases are due to inadequate hygiene, likely when the prepuce is difficult to retract or when phimosis is present. Balano-posthitis may cause discharge and urinary symptoms and is frequently irritant. Mild cases show diffuse, patchy or generalised erythema with scanty exudate. More severe cases show erosions which may become secondarily infected, and can cause profuse discharge. Occasionally painful inguinal lymphadenitis is found. Diabetes mellitus may present with balano-posthitis. Routine testing for glycosuria is essential in all cases.

Plasma cell balanitis of Zoon
This is a rare cause of chronic balanitis usually seen in the middle-aged or elderly. Examination shows one or more superficial red moist shiny plaques with central stippling reminiscent of 'Cayenne pepper'. Clinically the lesion is similar to the pre-malignant erythroplastic conditions (see p. 301) but is benign. Diagnosis is established by biopsy, which shows characteristic plasma cell infiltration.

Aetiology of balano-posthitis

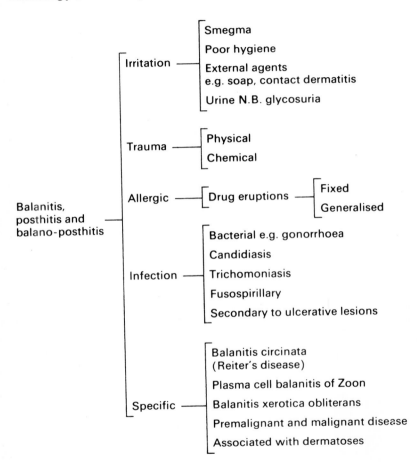

Balanitis, posthitis and balano-posthitis

- Irritation
  - Smegma
  - Poor hygiene
  - External agents e.g. soap, contact dermatitis
  - Urine N.B. glycosuria
- Trauma
  - Physical
  - Chemical
- Allergic
  - Drug eruptions
    - Fixed
    - Generalised
- Infection
  - Bacterial e.g. gonorrhoea
  - Candidiasis
  - Trichomoniasis
  - Fusospirillary
  - Secondary to ulcerative lesions
- Specific
  - Balanitis circinata (Reiter's disease)
  - Plasma cell balanitis of Zoon
  - Balanitis xerotica obliterans
  - Premalignant and malignant disease
  - Associated with dermatoses

365 Mild superficial balano-posthitis.
366 Plasma cell balanitis of Zoon. The appearance is non-specific; the diagnosis is determined by histology.
367 Severe balano-posthitis with phimosis and multiple erosions.
368 Chronic diffuse balanitis. Note the fibrotic prepuce.
369 Patchy balanitis associated with preputial adhesions.
370 Patchy erosive balanitis: these lesions are usually secondarily infected.

365

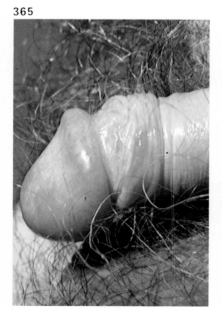

366

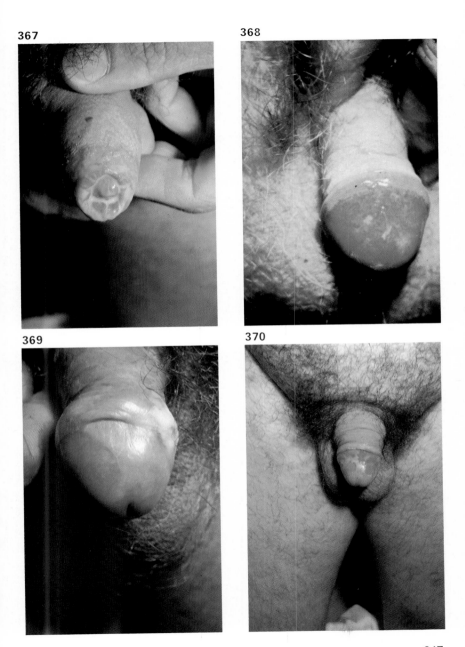

367

368

369

370

371 Balano-posthitis, mainly affecting coronal sulcus.
372 & 373 Ecthyma of glans and prepuce. Note irregular infected ulcerated crusted lesions which healed with minimal scarring.

371

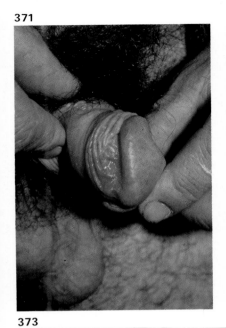

372

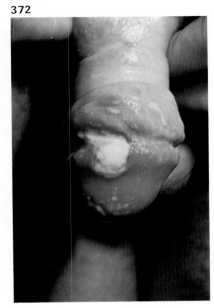

373

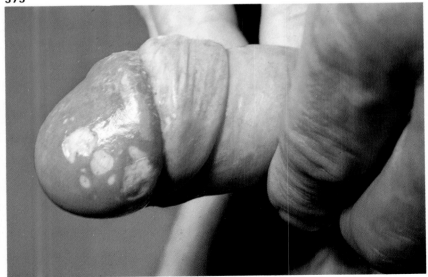

BEHÇET'S SYNDROME

Genital aphthosis

The term *genital aphthosis* is unsatisfactory because it has no adequate definition. It is used to include recurrent minute genital erosions, *Lipschütz ulcers* of the vulva and *Behçets syndrome*, but it is not clear whether these are separate entities or differing manifestations of a common, unknown cause. It seems probable that a proportion of cases diagnosed as genital aphthous ulceration and Lipschütz ulcers in the past were due to unrecognised *herpesvirus hominis* infection.

Behçet's syndrome (triple syndrome)

Behçet's syndrome is a condition characterised by recurring *genital* and *oral ulceration,* frequently associated with *eye lesions* and *pyoderma* and, occasionally, in late stages, with development of neurological, gastro-intestinal, cardiac or pulmonary lesions. *Formes frustes* are probably more common than is generally recognised, particularly as the course of the disease may be prolonged. The first manifestation is usually the appearance of morphologically similar ulcers on the genitalia and in the mouth, but lesions may not occur concurrently. The ulcers are usually very painful, 5–20 mm in diameter, with an erythematous halo and have a base covered with yellow slough. Single lesions are most common and scarring from previous lesions may be evident. In male patients the genital ulcer is usually found on the *scrotum* but may occur on the *penis*; in female patients ulceration is usually on the *labium majorum.* Oral lesions occur most commonly inside the *lower lip* but may be found anywhere within the buccal cavity. Ulcerated lesions usually heal in one or two weeks but occasionally persist longer. Eye symptoms include *conjunctivitis, uveitis* and *hypopyon.* The interval between relapses frequently exceeds a year and is very variable. The diagnosis is made on a basis of history and physical findings and exclusion of other conditions; there are no pathognomonic investigations.

374 Behçet's syndrome: iritis and hypopyon.
375 Behçet's syndrome: typical penile ulcer. Note erythematous margin.
376 Behçet's syndrome: typical scrotal ulceration. Compare with figs. 379 and 381.
377 Behçet's syndrome: solitary ulcer of scrotum.

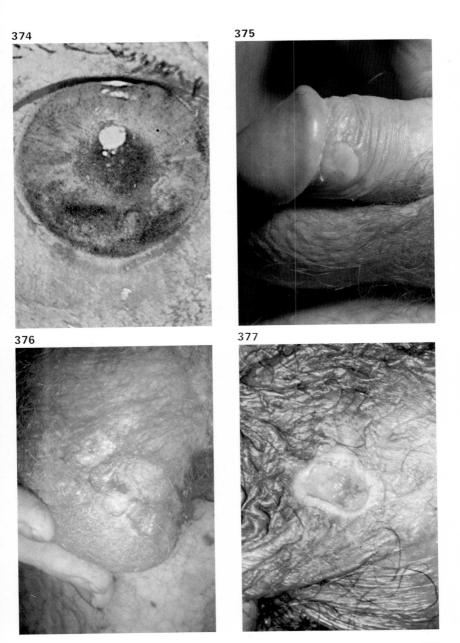

378 Behçet's syndrome: lip ulcer. Note scarring from previous episodes.
379 Behçet's syndrome: same patient as fig. 378: concurrent ulceration of labium minus.
380 Behçet's syndrome: lip ulcer. Note erythematous margin.
381 Behçet's syndrome: multiple ulcers of vulva.

378

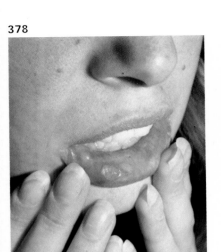

379

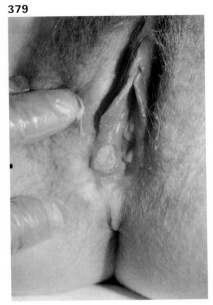

380

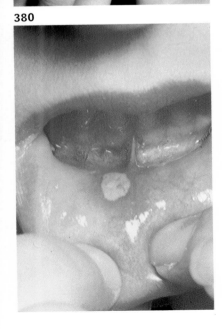

381

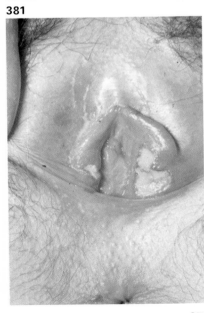

ERYTHRASMA

Erythrasma of the groin is a relatively common, unimportant, chronic cutaneous infection. It is seldom the sole reason for attendance and is usually found fortuitously. It is caused by *Corynebacterium minutissimum* which can be demonstrated in stained smears taken from involved areas. Infection commonly occurs in skin folds; in the anogenital area it may be found on the pubis, scrotum, thigh, groin and natal cleft. Lesions are usually well-defined, pink or brown in colour and dry and scaly. Mild irritation may occur but often the condition is asymptomatic. Examination of affected areas under Wood's light shows a characteristic coral-pink fluorescence.

382 Erythrasma of groin. Compare with tinea cruris (fig. 390) and candidiasis (fig. 352).
383 Erythrasma of axilla.
384 Erythrasma of scrotum and thigh.

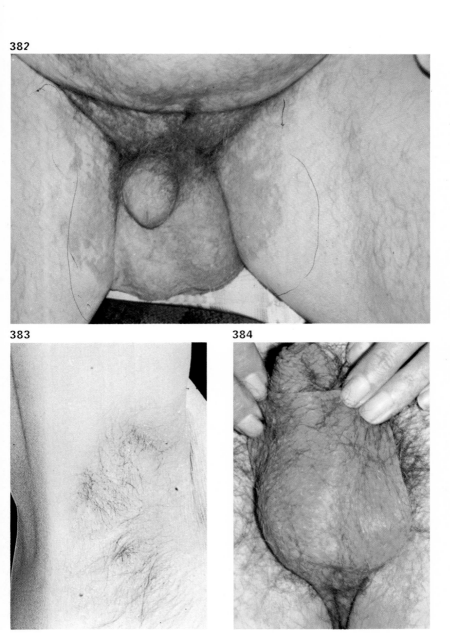

382

383

384

HIDROADENITIS

Apocrine glands in the anogenital area are found in the groin, and in the scrotal, vulval and perianal regions. These glands may be the site of a chronic infection which may eventually result in considerable tissue damage. The disease usually begins with discrete small subcutaneous nodules which break down to form abscesses that usually later become confluent and discharge on the surface through one or several sinuses. Active lesions and scarring from previous lesions are usually observed. In venereological practice the condition is rare.

385 Hidroadenitis. Note old and active sinuses and evidence of subcutaneous inflammatory lesions.

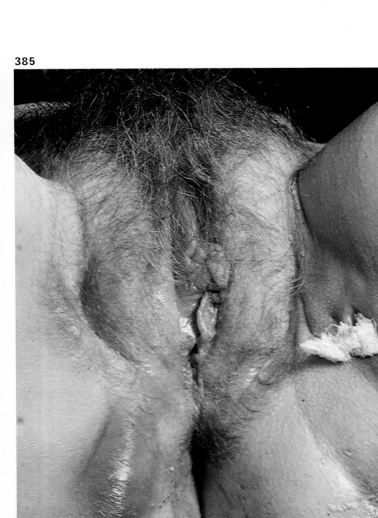

OTHER CONDITIONS

FUNGAL INFECTIONS

Fungal infections (other than candidiasis, see p. 222) of the anogenital area are common and may affect the penis, groin, scrotum, thigh, vulva or perianal regions: males are more frequently affected than females. The lesion on the thigh and scrotum is known as *tinea cruris ('Dhobie itch')*; in many cases tinea pedis can also be found. The diagnosis is established by finding fungal elements in KOH preparations (see p. 70) or by culture. *Epidermophyton floccosum* and *Trichophyton rubrum* are the organisms most often found. The lesion is usually moderately irritant and spreads slowly. The margin of the area is red and often small satellite lesions are present. Centrally, the involved skin is less erythematous and scaly. Vesiculation occasionally occurs but is rare.

Tinea versicolor, caused by *Microsporon furfur*, is another fungal condition that commonly affects the upper parts of the body but may also affect the genitalia. The infection is most common in coloured patients. Discrete round macules, 1–10 mm in diameter, may be found on the lesions and may become confluent and involve considerable areas. The lesions are *café-au-lait* colour. Diagnosis is established by mycology.

386 Fungal infection of glans penis and nails: compare with fig. 351.
387 Fungal infection of glans penis.
388 Fungal infection of glans penis. Note marginal activity. The patient later developed psoriasis.

386

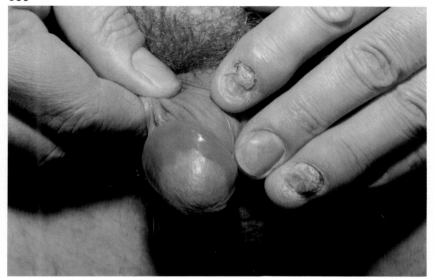

387

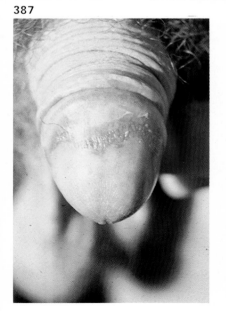

388

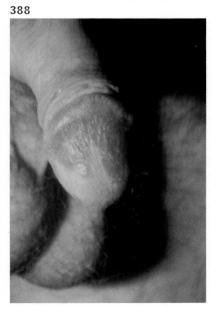

259

389 Tinea cruris: close-up of edge of lesion to show satellite follicular lesion.
390 Tinea cruris: typical lesion on thigh.
391 Tinea versicolor: note morphology and depigmentation.

389

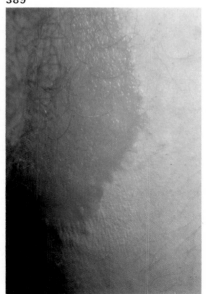

390

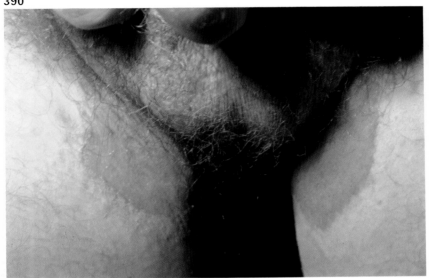

391

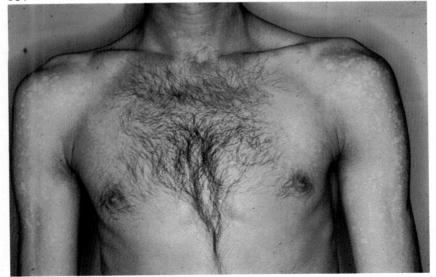

LICHEN SIMPLEX

Manifestations of lichen simplex (*neurodermatitis*) are common in the
anogenital region. The condition occurs in predisposed individuals and is
usually precipitated by a 'trigger' factor which may be physical or emotional.
Severe pruritus ensues and often persists even after the precipitant factor
has gone; psychogenic influences are marked. Lesions may be localised or
generalised and are often bilaterally symmetrical. Affected areas show
lichenification and some oedema, and excoriation, fissuring and hyperpigmentation may be present. Dusky erythema is usual, but red or white
areas may also be observed. Lesions may occur anywhere in the anogenital
region but are most common on the vulva (especially in the para-clitoral
area), the scrotum and in the groin.

392 Lichen simplex: severe involvement of pubis, genitalia and thigh.
393 Lichen simplex: anus.
394 Lichen simplex: early lesion on vulva.
395 Lichen simplex: advanced lesions of vulva and perianal region.

392

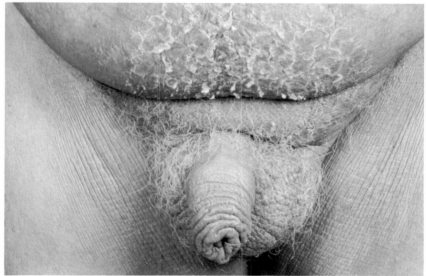

393

394

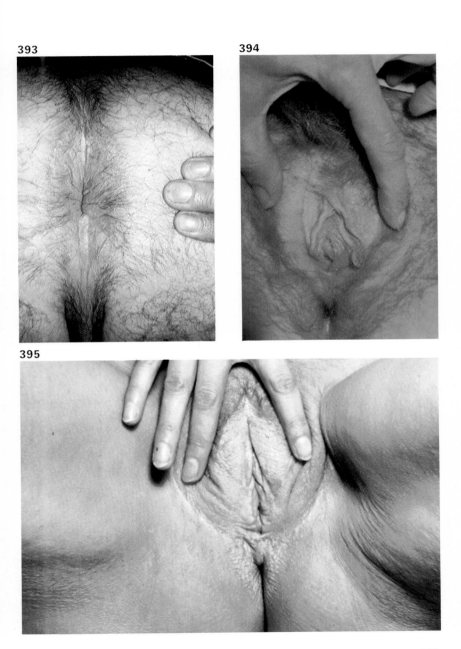

395

263

LICHEN PLANUS

Lichen planus may cause chronic irritating papular lesions in both males and females. The onset is insidious—the patients attend complaining of irritant spots or rash. The lesions are usually papular, 2–15 mm in diameter, sometimes grouped or coalescent. Lesions are usually dusky pink or violaceous in colour but may be white; the surface shows a network of fine lines (*Wickham's striae*). Lesions of lichen planus may be found anywhere on the genitalia but are most common on the shaft and glans of the penis. In the majority of cases other manifestations of lichen planus are present. Particularly characteristic is the lacy white network on the buccal mucosa of the cheek.

The cause of lichen planus is unknown but psychogenic factors play a considerable part. Lesions usually clear up in less than a year but recurrence occurs in 15 to 20 per cent of cases.

396 Lichen planus: typical violaceous papules on glans penis.
397 Lichen planus: ivory white lesions on glans penis and prepuce.
398 Lichen planus: glans penis.
399 Lichen planus: coalescent papules on glans penis.

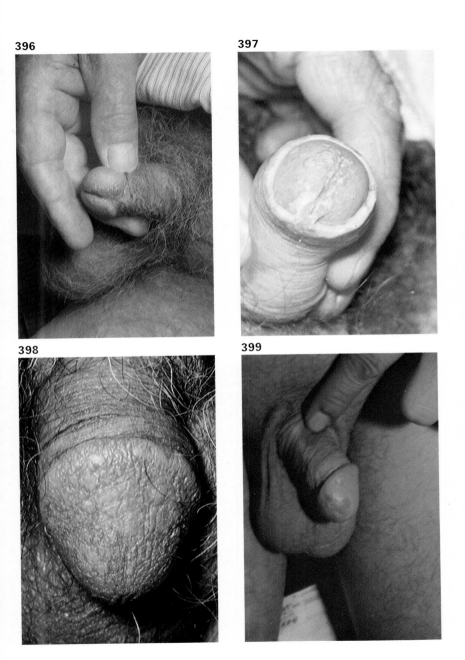

396

397

398

399

400 Lichen planus: lesions of penis and wrist.
401 Lichen planus: 'chinese white' patches inside cheek.
402 Lichen planus: typical lesions on penile shaft.
403 Lichen planus: same patient as fig. 402: glans penis.

400

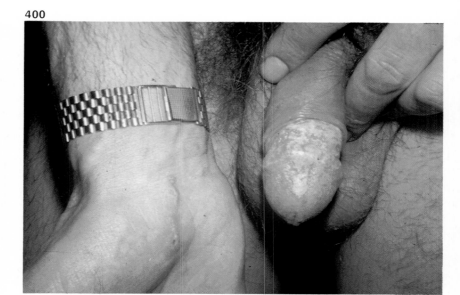

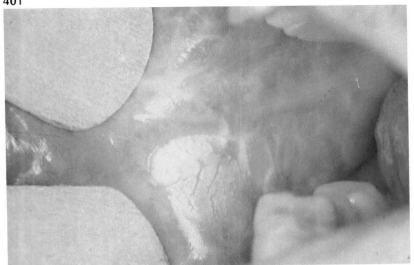

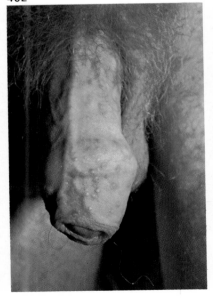

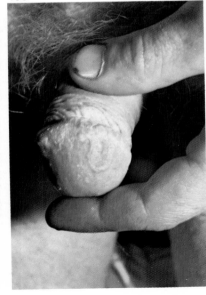

LICHEN NITIDUS

Lichen nitidus is thought to be a variant of lichen planus. The condition
is frequently asymptomatic but irritation may occur. The clinical appearance
is of tiny, white shiny papules, usually multiple and often grouped.
Lichen nitidus is seen most often on the penile shaft. Transient and persistent
lesions appear to be equally common.

404 Lichen planus: extensive lesion on glans penis and scrotum.
405 Lichen planus: chronic lesions on penile shaft.
406 Lichen planus: lesions on penile shaft: note active and healed
areas.
407 Lichen nitidus.

404

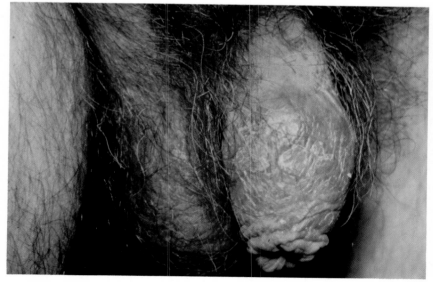

405

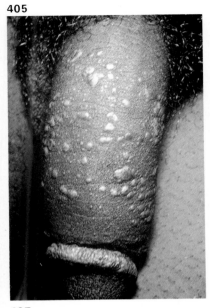

406

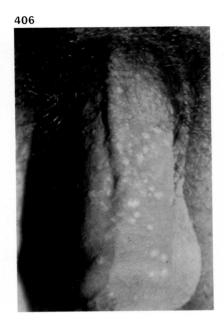

407

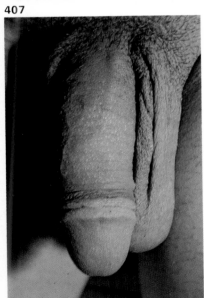

PEMPHIGUS VULGARIS

This is an uncommon chronic blistering disease, of unknown cause, which occurs more commonly in the middle-aged and in individuals of Jewish origin. The mucous membranes of the vulva and penis are occasionally involved, usually with evidence of the disease elsewhere on the skin or in the mouth. Bullous and erosive lesions may be found; on mucous membranes the bullous stage is often transient, and the erosive lesions painful. The lesions on other skin sites are often initially localised. After healing areas of hyperpigmentation may be found.

PSEUDOACANTHOSIS NIGRICANS

True acanthosis nigricans is associated with malignant disease. A morphologically similar lesion is occasionally seen in obese patients or may occur at puberty. The condition is asymptomatic; examination shows hyperpigmentation and slight lichenification of the vulva, upper thigh and perianal regions.

408 Pemphigus: compare with fig. 340, candidial balano-posthitis.
409 Pseudoacanthosis nigricans: marked perianal and moderate thigh pigmentation.

408

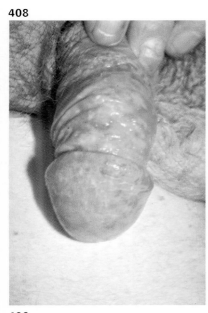

409

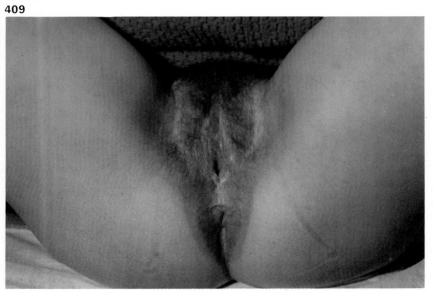

PITYRIASIS ROSEA

This condition of unknown aetiology is a classical misdiagnosis for secondary syphilis. The first sign of the disease is the appearance of a *herald patch,* which may sometimes occur on the penis but is more often found on the trunk or neck, and occasionally at other sites. The herald patch is a round or oval bright red solitary sharply defined lesion covered with fine scales; the lesion may be 2—4 cm in diameter. The general eruption appears in crops beginning one or two weeks after the herald patch. Multiple lesions appear which often follow the lines of cleavage and are commonly found on the trunk and proximal parts of the limbs. The oval or rounded scaly lesions are dull pink at the margin and show central clearing; the size seldom exceeds 1 cm diameter. Macules are occasionally seen. Irritation often occurs both in the herald patch and in the general eruption in contradistinction to the usually non-irritant secondary syphilis. The eruption clears without residual scarring in three to six weeks.

410 Pityriasis rosea: lesions on trunk. Note resemblance to maculo-papular secondary syphilide.
411 Pityriasis rosea: lesions of thigh and penis.

410

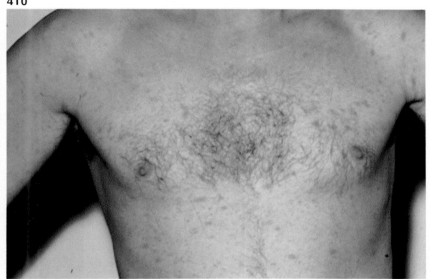

411

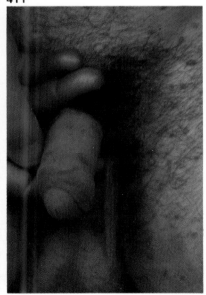

PSORIASIS

Genital lesions are frequently found in this common skin disease of unknown cause. Diagnosis is usually easy because of evidence of psoriasis elsewhere but genital lesions occasionally occur alone. In many patients with genital psoriasis typical circumscribed scaly lesions are found, but lesions of the vulva or of the glans penis in uncircumcised males may be considerably modified. Lesions in these situations often are non-scaly and differentiation from other causes of balanitis or vulvitis may be difficult. The lesions are usually bright red and sharply marginated: lichenification (especially in vulval lesions) may be severe. Psoriatic lesions are usually irritant. The course of the disease is marked by spontaneous remission and relapse.

412 & 413 Psoriasis of glans penis and prepuce. Note the colour and sharply demarcated margin.
414 Psoriasis of circumcised penis and nails. Note the resemblance of the penile lesion to lichen planus, and also compare with figs. 351 (candidiasis) and 386 (fungal infection).

412

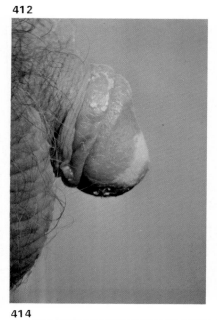

413

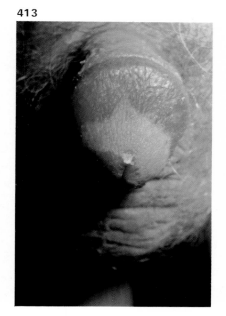

414

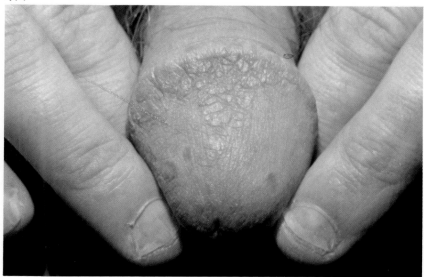

415 Psoriasis: guttate and flexural lesions.
416 Psoriasis: flexural lesions with sharply demarcated edges. Compare with fig. 328, chronic candidal intertrigo.

415

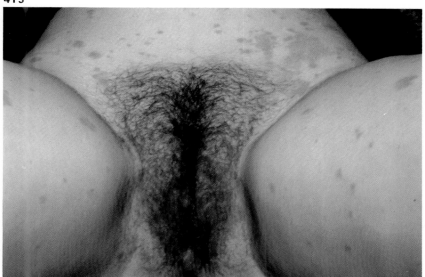

416

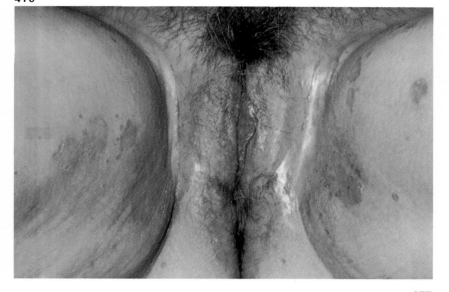

SEBORRHOEIC DERMATITIS

The anogenital area is a common site for manifestations of the 'seborrhoeic state' in predisposed individuals. Lesions may be found in pubic, crural, vulval and perianal regions. Usually there is evidence of seborrhoeic dermatitis elsewhere on the body but if anogenital lesions occur alone differentiation from candidiasis, tinea or psoriasis can be extremely difficult. The lesions are yellowish-red or dull red in colour, diffuse and covered with greasy scales. Chronic lesions may show eczematous change. In flexural sites the appearance is of intertrigo, often with crusted fissures and secondary infection. Mild to moderate pruritus is common.

417 Seborrhoeic dermatitis of groin, showing intertrigo and folliculitis.
418 Seborrhoeic dermatitis of thigh and scrotum. Note the yellowish scaling. *Candida albicans* was grown on culture but the patient had unequivocal seborrhoeic dermatitis of the chest.

417

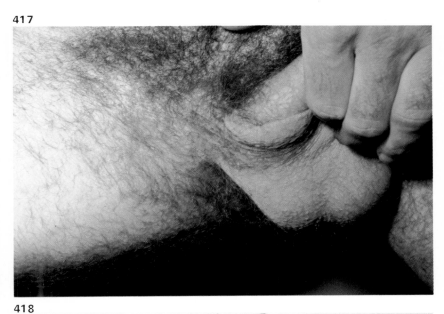

418

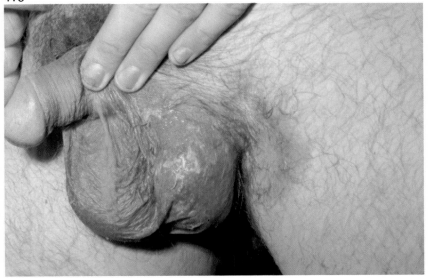

DRUG REACTIONS

Adverse reactions to drugs include toxic effects (which are beyond the scope of this atlas), contact dermatitis (discussed on p. 310) and allergy. Allergic reactions may be either *localised (fixed drug eruption)* or *generalised*. The range of drugs that can produce allergic reactions is enormous but in practice relatively few drugs produce most of the reactions observed. In this section common manifestations of drug allergy seen in venereological practice are discussed.

Fixed drug eruptions

Fixed drug eruptions occasionally occur in the anogenital region but the reason for this selectivity is unknown. The majority of lesions seen in practice are caused by barbiturates, sulphonamides or other antibiotics. The reaction usually appears shortly after exposure to the allergenic drug as a single (occasionally multiple) well-defined oedematous or vesicular area 5 to 20 mm in diameter, dusky red or brown in colour. The lesion is often extremely irritating and pigmentary changes may persist for a considerable time. The history of a lesion which reappears in identical form each time a particular drug is used is diagnostic.

Generalised drug reactions

Generalised drug reactions have protean manifestations and the reader is referred to textbooks of dermatology for comprehensive description. From the standpoint of the venereologist the drugs most commonly seen to cause adverse reactions are penicillin and other antibiotics, barbiturates, and anti-inflammatory agents such as the salicylates and phenylbutazone and related compounds. The types of allergic reaction most frequently observed are urticaria, exanthemata, exfoliative lesions, serum sickness and the Stevens–Johnson syndrome (see p. 342).

419 Fixed drug eruptions: tetracycline.
420 Fixed drug eruptions: penicillin. The lesion simulates balano-posthitis. Note also hirsutes follicularis.
421 Fixed drug eruptions: barbiturate.
422 Fixed drug eruptions: barbiturate.

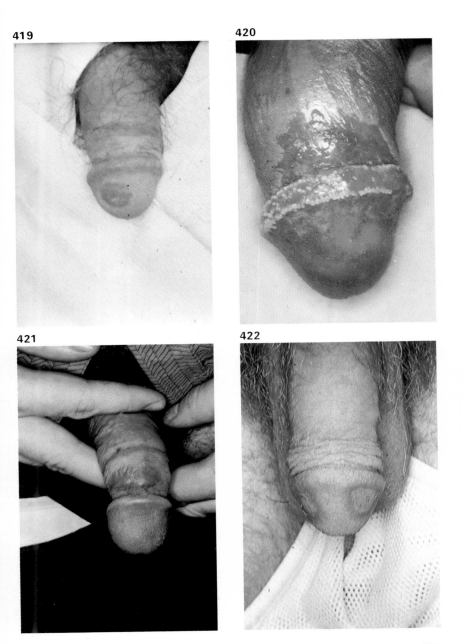

419

420

421

422

423 Fixed drug eruptions: further examples.
424 Fixed drug eruptions: further examples.
425 Fixed drug eruptions: further examples.
426 Generalised drug reactions: phenylbutazone. Lesions on hands.
427 Generalised drug reactions: sulphonamide reaction simulating secondary syphilis.

423

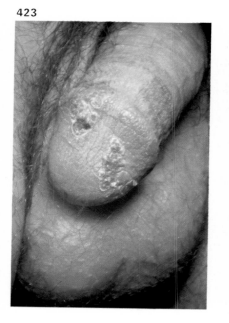

424

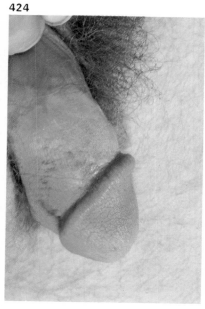

425

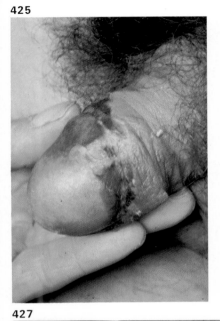

426

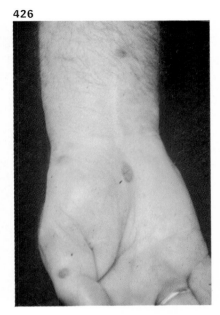

427

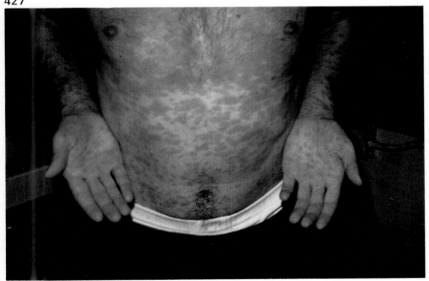

428 Generalised drug reactions: phenylbutazone. Macular rash of chest.
429 Generalised drug reactions: penicillin. Scarlatiniform reaction.

428
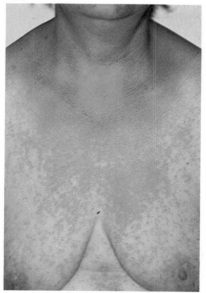

429
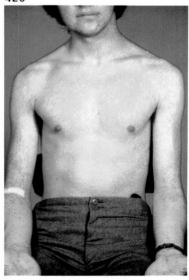

DYSTROPHIC CONDITIONS

There is considerable controversy concerning the relationship of morphologically similar lesions variously labelled *kraurosis, leukoplakia, primary
vulval atrophy* and *lichen sclerosus et atrophicus (LS & A). Balanitis xerotica
obliterans (BXO)* is thought to be the equivalent in male patients of LS & A.
A considerable number of authorities think that LS & A is the common
underlying condition for patients with dystrophic lesions. LS & A affects
both glabrous and hairy surfaces; primary vulval atrophy affects hairy
surfaces only. The cause of both conditions is unknown.

Lichen sclerosus et atrophicus (LS & A)

LS & A is a generalised skin disease with frequent genital manifestations.
It may occur in both males and females; the condition in males (*BXO*) is
described later. Genital lesions in women are found most frequently in
middle age but are not uncommon in younger and older age-groups. Some
30 to 40 per cent of women with vulval lesions have evidence of the disease
in other areas. Vulval lesions may be asymptomatic but are usually
irritant or may present with dyspareunia due to vulval atrophy. Initial vulval
lesions are papular and may appear either pale or erythematous. Hyperkeratosis and telangiectasia are often seen. Lesions may first appear on or
near the clitoris but can be anywhere on the vulva. Progress is irregular but
eventually the whole of the vulva and the anal and perianal regions may be
involved. In advanced cases the skin is pale, thin and atrophic. Similar
anal and perianal lesions may be seen in male patients. When genital lesions
are found evidence of the disease should be sought in other areas, particularly
the upper trunk, neck and axillae. Lesions in these situations are slightly
raised, flat, white macules or papules, often grouped.

Balanitis xerotica obliterans (BXO)

A considerable proportion of patients with the penile lesions of BXO are
found to have lesions of LS & A on other parts of the body. BXO alone is a
common condition. Many patients remain asymptomatic for a long time.
Malignant change has been reported. The condition frequently begins in
childhood.

In uncircumcised patients pale, fibrotic areas are found at the preputial
margin and glans penis. Lesions are often macular and with progress
phimosis, recurrent fissuring of the prepuce and adhesion of the prepuce
to the glans may occur. Lesions of the glans are similar to those seen in
circumcised patients.

In circumcised patients localised pallid areas with diffuse margins are

285

seen on the glans, more commonly on the ventral surface. In early cases lesions may be erythematous. Telangiectases may occur in or at the margin of affected areas and bleeding occasionally occurs. In some patients the fibrosis is confined to the perimeatal region, resulting in meatal narrowing (stenosis) and stricture formation. Patients with meatal stenosis may present as cases of non-gonococcal urethritis.

430 Lichen sclerosus et atrophicus: early periurethral papules with atrophic areas of vestibule.
431 LS & A: extensive vulval lesions.
432 LS & A: vulval atrophy. Note minute telangiectases.
433 LS & A: extensive perivulval and perianal atrophy.
434 LS & A: perianal lesion.
435 LS & A: grouped papules in axilla.

430

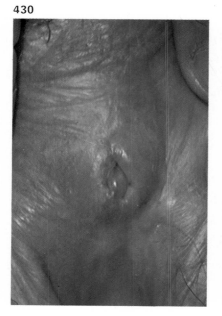

431

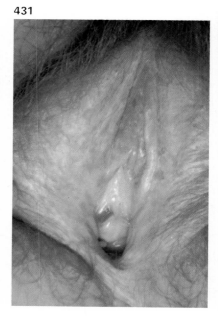

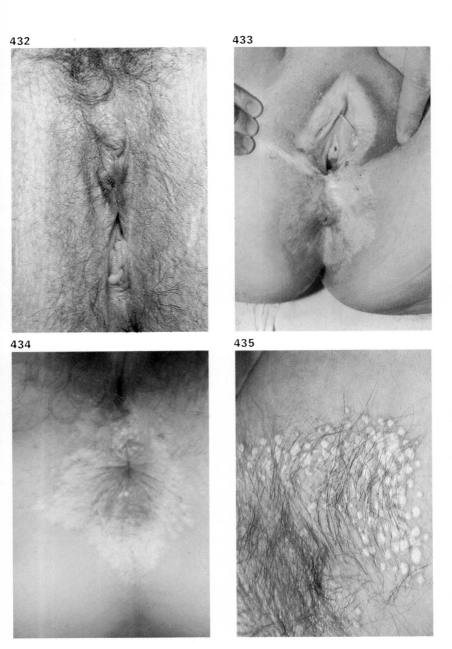

432

433

434

435

436 Balanitis xerotica obliterans: early erythematous lesion in uncircumcised patient.
437 BXO: perimeatal atrophy and telangiectasia.
438 BXO: lesions of glans penis, coronal sulcus and prepuce.
439 BXO: meatal stenosis.
440 BXO: extensive lesions of shaft and glans penis; prepuce is also affected.
441 BXO: extensive lesions on glans penis. Same patient as fig. 440.

436

437

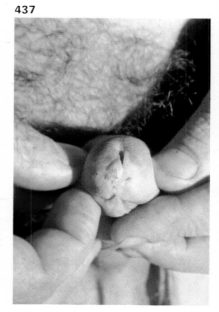

438

439

440

441

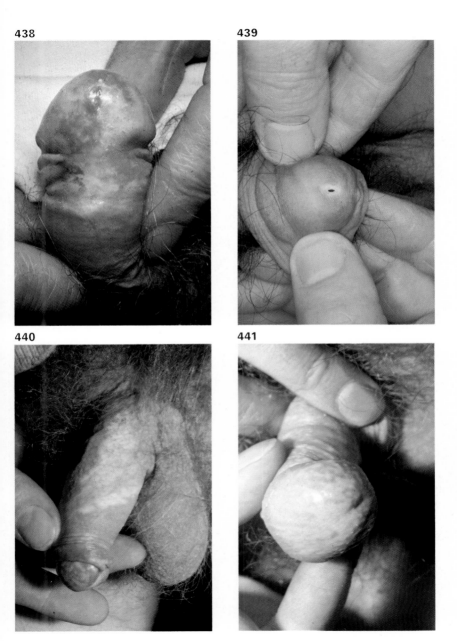

442 BXO: phimosis. Note pallor of glans.
443 BXO: phimosis. Note ring of fibrosis.

442

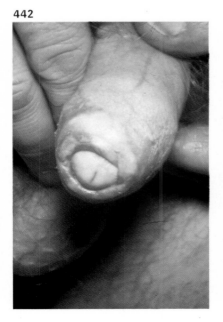

443

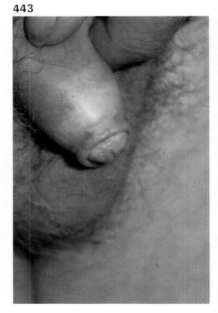

PARASITIC INFESTATIONS

Pediculosis pubis

Pubic lice (*Phthirus pubis, crabs*) are commonly transmitted during sexual contact but may also be acquired from infected fomites such as shed hairs, clothing or towels. The lice feed on the skin surface and intense irritation is produced at the sites of bites. The lice and their eggs (*nits*) are found on hairy locations such as the pubis, perineum, buttocks and upper thighs; occasionally they are found in the axillae and rarely on the eyelashes. Crab lice live for about four weeks. The female lays 8–10 eggs daily which hatch in about eight days. Maturity is reached in a further week.

The patient may become aware of infestation because of irritation or by noticing movement of the small (1–2 mm) yellow-brown or grey lice or by finding the minute black or dark brown nits attached to the base of a hair.

Occasionally the first evidence of infestation is the appearance of pinhead-sized blood spots on the underwear. Excoriation and secondary infection of affected areas is common. The diagnosis is made by recognition of the louse or nit.

It is worth remembering that parasitophobia is not uncommon: definite evidence of infestation should be sought before treatment is given. A self-inflicted hazard of louse infestation is the severe dermatitis that can be caused by mercury-containing ointments which are traditionally used in treatment (see fig. 491).

444 Pediculosis: infestation of pubis, thighs and scrotum with secondary folliculitis.
445 Pediculosis: nit containing larva attached to hair.
446 Pediculosis: adult louse attached to hair.
447 Pediculosis: nits attached to eyelashes.
448 Pediculosis: close-up view showing lice and nits on pubis.

444

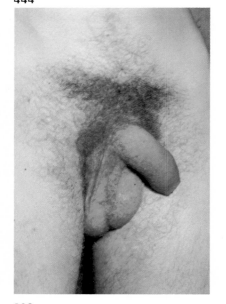

445

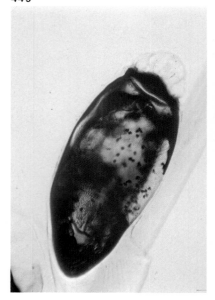

446

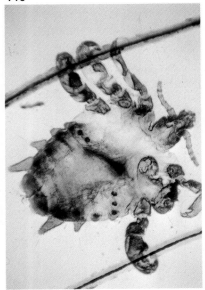

447

448

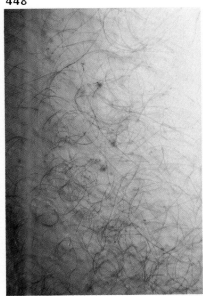

Scabies

The mite *Sarcoptes scabiei* is frequently transmitted during sexual contact, producing characteristic genital lesions. The lesions may be found anywhere on the genitalia and are also frequently found on the abdomen, wrist, hand and elbow; occasionally other locations may be infested but the face is spared. The lesions are intensely irritant and, characteristically, irritation is exacerbated at night or when the patient is warm.

The life of the mite is about three weeks. After hatching in a burrow the larvae migrate to a skin pocket and moult several times: maturity is reached in about two weeks. The mature mite burrows into the stratum corneum, advancing 1–2 mm daily. The female mite lays two or three eggs daily which hatch in three or four days.

The classical clinical lesion is a slightly raised sinuous burrow, which is often most readily identified between the fingers. At other sites excoriation, secondary infection or eczematous change may distort the appearance. Lesions on the penis and scrotum are frequently oedematous, and lesions on infants may be vesicular.

Diagnosis is usually based on the history and findings: mites can sometimes be extracted from burrows and recognised.

449 Scabies: adult *Sarcoptes scabiei* (microscopic view).
450 Scabies: typical extensive scabies of penis, thighs and abdomen.

449

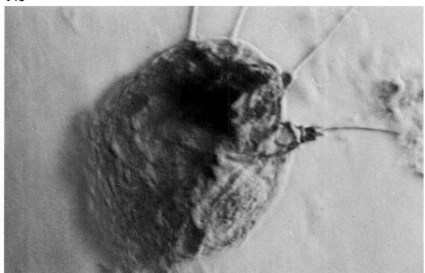

450

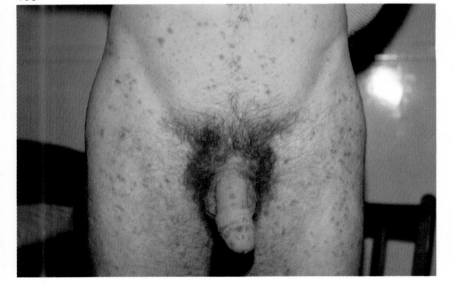

451 Scabies: lesions on glans penis; close-up view of fig. 450.
452 Scabies: oedematous lesions on penile shaft.
453 Scabies: secondarily infected scabies with inguinal bubo. Note typical oedematous lesions on scrotum and penis. The bleeding lesion on the prepuce was scarified for preparation of dark-ground specimens.
454 Scabies: burrow on penis.
455 Scabies: interdigital burrow.

451

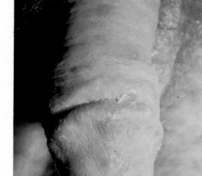

452

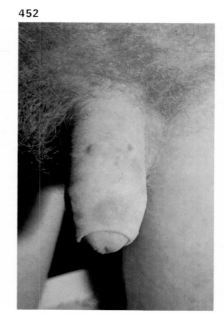

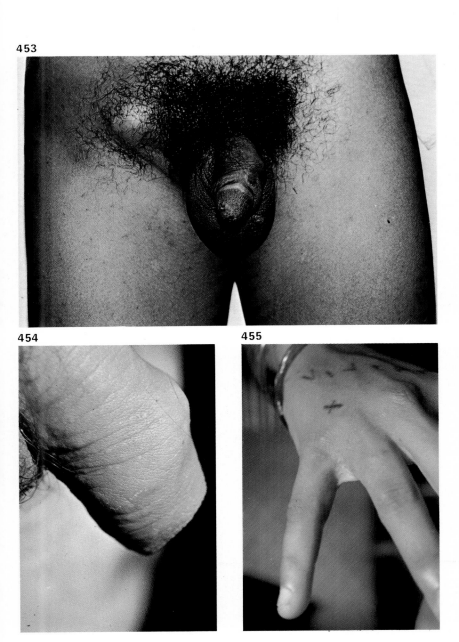

453

454

455

456 Scabies: eczematised lesions.
457 Scabies: vesicular lesions in infant.

456

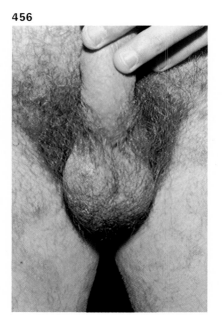

457

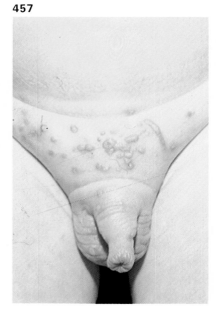

PRE-MALIGNANT AND MALIGNANT CONDITIONS

Pre-malignant and malignant conditions occasionally occur in venereological practice but are rare. *Erythroplasia of Queyrat, Bowen's disease, Paget's disease, basal-cell* and *squamous-cell carcinomata* and other neoplasms may be found in the anogenital regions. Reticuloses or secondary deposits may involve inguinal lymph glands. In advanced cases the diagnosis of malignant disease is usually obvious but early stages may present as lesions simulating balanitis or vulvitis or as genital ulcers. Clinical differentiation is usually impossible and the diagnosis has to be established by biopsy.

Erythroplasia of Queyrat

This pre-malignant condition may be found on the penis (particularly in uncircumcised patients) or on the vulva. Examination shows a slightly elevated plaque, sharply demarcated with a bright red, clean, velvety surface. The lesion spreads slowly.

Bowen's disease

This intra-epidermal carcinoma may be associated with systemic malignant disease. Lesions are found on the penis (frequently in association with chronic balano-posthitis) or vulva. Examination shows a non-elevated plaque with an irregular margin: the lesion has a dark red crusted or scaly surface. Pruritus of the patch is frequent, and the lesion spreads slowly.

Paget's disease

This intra-epithelial carcinoma is frequently associated with an underlying carcinoma and it has been suggested that the surface lesion is a secondary growth. Lesions may be found on the vulva and perianal regions and occasionally on the penis. Examination shows a circumscribed scaly and crusted erythematous plaque, sometimes ulcerated and oozing blood. The lesion is often irritant.

Basal-cell carcinoma (epithelioma, rodent ulcer)

Basal-cell carcinoma is much less common than squamous-cell growths. Typical 'rodent ulcers' with a pearly, firm, rolled edge are most often found on the penis: they are rare in other genital areas. 'Chimney sweep's cancer' (epithelioma of the scrotum), an occupational hazard, is now very uncommon.

458 Erythroplasia of Queyrat. Lesion showing typical colour, surface and margin.
459 Erythroplasia of Queyrat. Atypical lesion showing peripheral changes due to chronic balano-posthitis. Diagnosis established by biopsy.
460 Erythroplasia of Queyrat: early lesion.
461 Epithelioma of penis (basal-cell carcinoma). Note typical rolled edge.
462 Squamous-cell carcinoma: extensive superficial warty lesion. The patient had noticed the lesion only four weeks previously.

458

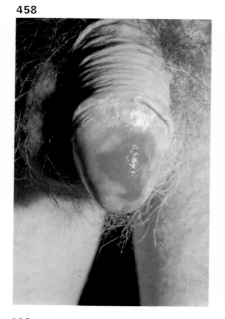

459

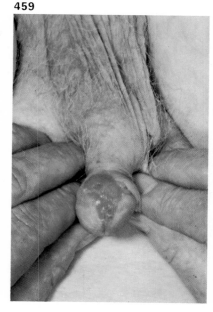

460

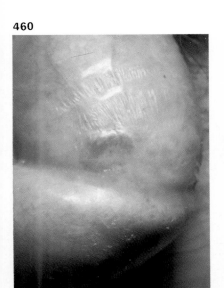

461

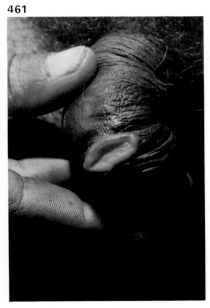

462

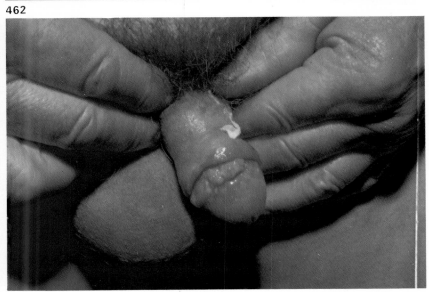

Squamous-cell carcinoma

This condition usually develops after middle age and is most common in uncircumcised males. Attendance is often delayed because the lesion is painless. It may present as a warty growth or as an indurated ulcer with a firm edge. Spread is often rapid and involvement of inguinal lymph nodes is frequent. Growths arising from urethral epithelium may present with lumps in the penis, urethral discharge or urinary obstruction.

463 Squamous-cell carcinoma: warty lesion.
464 Squamous-cell carcinoma: ulcerated lesion.
465 Carcinoma of penis: origin unknown. The patient presented as Peyronie's disease and the condition remained unchanged for two years. The true diagnosis was established after spread to inguinal nodes and biopsy.
466 Neoplastic papilloma of penis.
467 Advanced squamous cell carcinoma of penis with secondary deposits in the inguinal glands. The patient was aged 42 and presented with anxiety that he had a venereal disease. Symptoms had been present for one year.

463

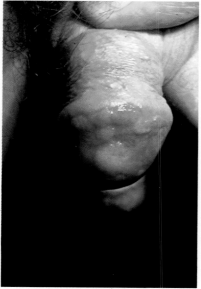

464

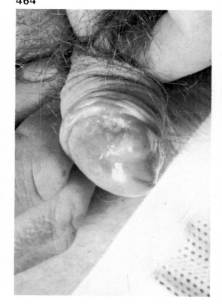

465

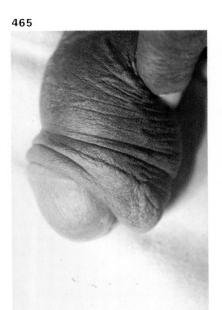

466

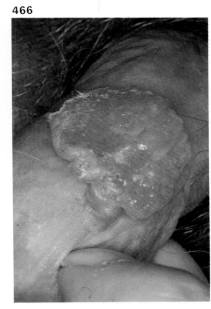

467

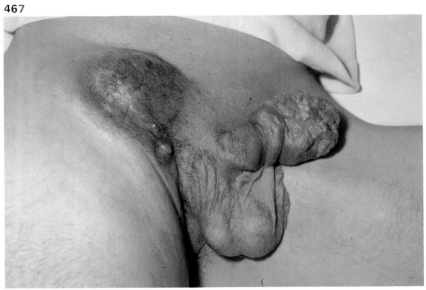

468 Squamous cell carcinoma with almost complete destruction of the penis and with fungating secondary deposits. Patient was aged 72 and referred as 'primary syphilis'.

468

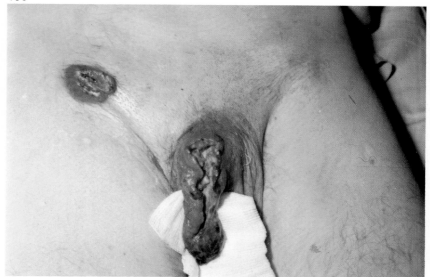

PYOGENIC LESIONS

The anogenital region, in common with all areas of the body, may be the site of pyogenic skin infections, whose appearance may arouse fears of sexually transmitted disease. Skin infections are particularly prone to occur in the area for anatomical reasons: hairy skin, flexures, numerous glandular structures and proximity to the anus. Septic penile lesions may cause gross local oedema, and adhesions between the prepuce and glans penis are likely to form pockets where infection can readily occur. *Folliculitis, furunculosis, abscess* and *ecthyma* may be found. The majority of infections are caused by *staphylococci* or *streptococci*. Fusospirillary infections may cause severe tissue destruction (*phagedena*) or gangrene.

469 Pubic folliculitis. Care must be taken to exclude parasitic infestations.
470 Napkin rash. Can be confused with congenital syphilis.
471 Scrotal folliculitis. Compare with fig. 556, leprosy.
472 'Saxophone penis'. Marked oedema associated with minor pyogenic infection.
473 Penile abscess: note similarity to shaft chancre of primary syphilis (fig. 71).
474 Infected cyst of glans penis.

469

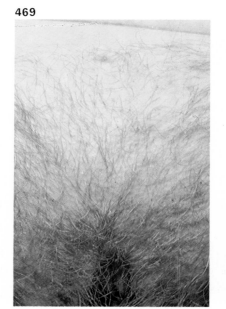

470

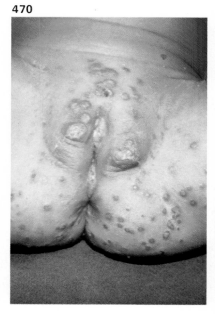

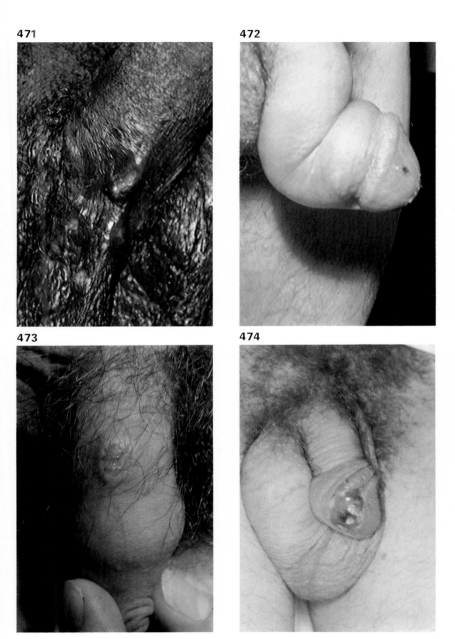

471

472

473

474

307

475 Penile abscess. Marked adjacent cellulitis and oedema.
476 Abscess arising in a coronal sulcus bridged by adhesions.
477 Secondarily infected herpes of penis.
478 Same patient as shown in fig. 477, showing the associated bilateral inguinal lymphadenitis.
479 & 480 Phagedena of penis. Fusospirillary infection, showing marked tissue destruction.

475

476

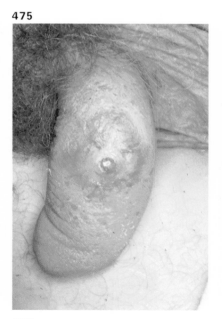
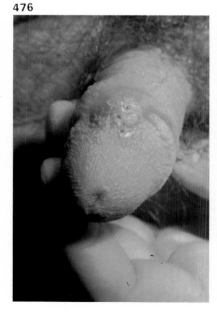

477

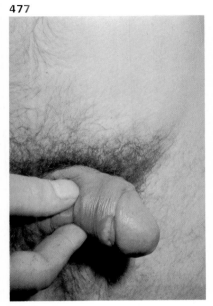

478

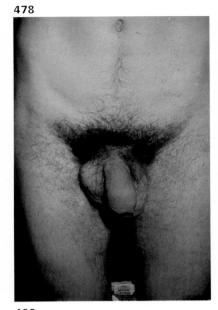

479

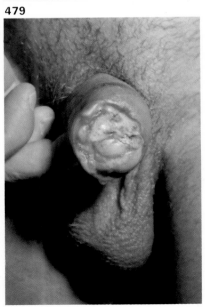

480

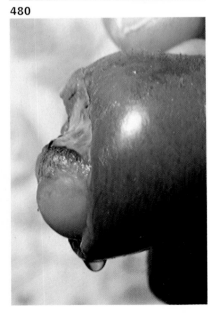

TRAUMA

Traumatic lesions of the genitalia due to physical causes are extremely diverse: the morphology is variable but history is usually diagnostic. Genital injury may occur during sexual intercourse or masturbation and is sometimes a feature of psychiatric illness. Bizarre sexual practices can result in bizarre physical lesions.

Drugs, disinfectants and other agents may act as irritants when in contact with the skin or mucous membranes of the genitalia in susceptible individuals and may cause *contact dermatitis*. The history is often diagnostic but patients may be reluctant to disclose the truth when the agent has been self-administered. Antiseptics such as 'Dettol' or potassium permanganate may be self-applied with prophylaxis against venereal disease in mind. Contact dermatitis may also be provoked by clothing or cosmetics; intravaginal contraceptive tablets, condoms and vaginal diaphragms are other occasional causes. The clinical findings are very variable and include pruritus, erythema, lichenification and, rarely, vesiculation. When a specific agent is suspected of causing contact dermatitis patch testing may be helpful.

Therapeutic measures (particularly radiation therapy) may result in lesions which can be termed traumatic.

481 Linear scarring resulting from zip-fastener injury. This cause more commonly produces ventral lesions.
482 Teeth marks after fellatio.
483 Penile abrasion.
484 Balano-posthitis after application of phenol to warts.

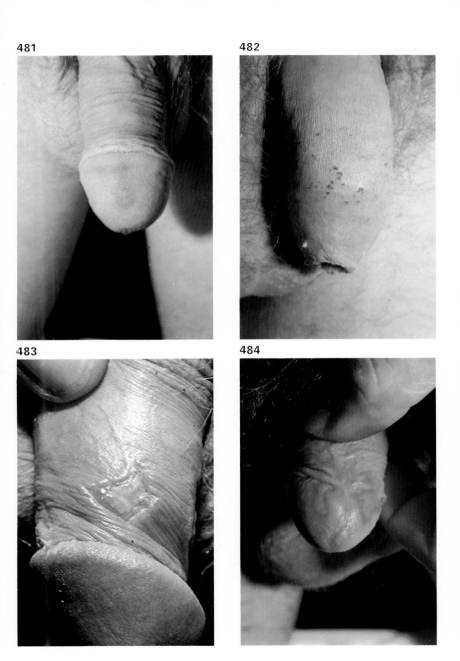

481

482

483

484

485 Contact dermatitis due to elastic in underwear.
486 Severe reaction to podophyllin ointment which was left on the skin overnight.
487 Penile and scrotal oedema due to use of vaginal deodorant!
488 Penile and scrotal contact dermatitis due to the use of a detergent containing enzymes.
489 Dermatitis artefacta. A bizarre lesion somewhat suggestive of lichen simplex.
490 Ecthyma after 'Id' reaction to fungus infection.

485

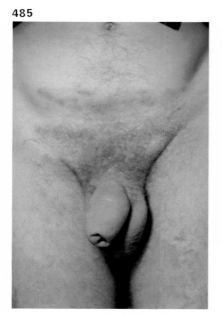

486

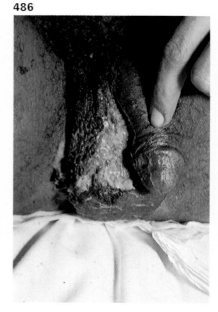

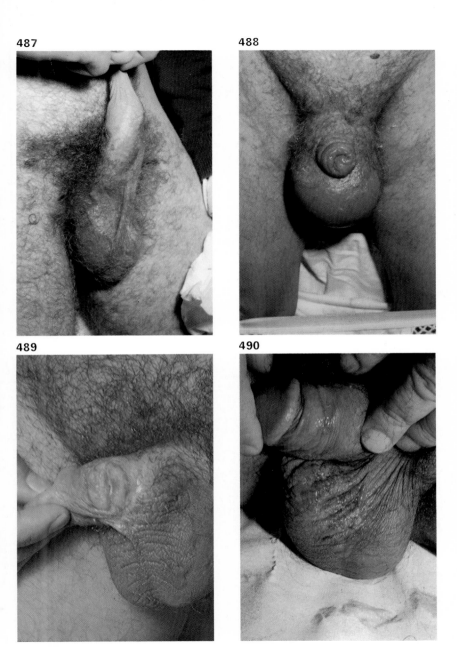

487

488

489

490

491 Contact dermatitis due to mercury ointment: note the characteristic colour.
492 Erosion of vaginal vault after douching with detergent (N.B. the patient is menstruating).
493 Vulvitis due to 'Dettol', which was diluted 1:2 only.
494 Post-radiation lymphangiectasis of pubis.
495 Post-radiation dermatitis of perianal region.

491

492

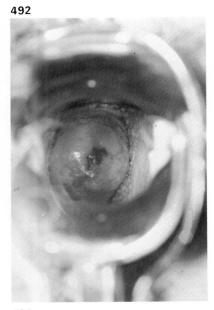

493

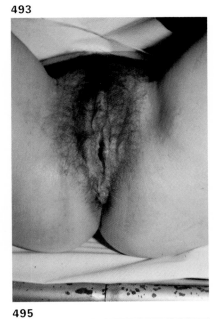

494

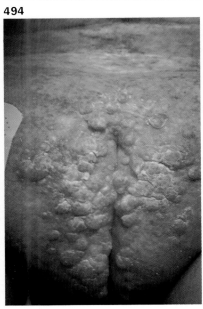

495

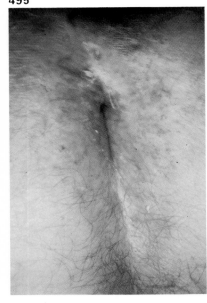

315

VIRUS DISEASES

Herpes genitalis (herpes simplex)

Herpetic lesions are the most common form of genital ulceration. Initial infection is frequently followed by recurrent attacks at irregular intervals, often over a period of several years. The infection may be transmitted accidentally or sexually (sometimes by oro-genital contact) in the first place but recurrences are not necessarily related to sexual activity. Transmission to sexual partners is only seen occasionally: perhaps this is because the majority of adults have herpes antibody in the serum (even in the absence of a history of the condition) and consequent immunity. There appears to be an association in females between genital herpes and the subsequent development of carcinoma of the cervix.

Cause
The virus *herpesvirus hominis type 2* is responsible for the majority of genital lesions: occasionally *herpesvirus hominis type 1* and *herpes zoster* may cause typical genital lesions. Virus is probably dormant in the tissues for long periods in many patients. Recrudescence may be caused by a variety of factors.

Clinical course
The incubation period of first attacks is short, usually three to six days. First attacks frequently occur in childhood and are often unrecognised. In recurrent attacks provocation often causes activity within one or two days.

The attack usually begins with the appearance of a group (or groups) of tiny papules, which may be irritant or painful. Irritation may occur at the site of eruption before the lesions appear. The papules develop into small, clear or yellowish vesicles. The surface of the vesicle erodes (particularly at sites subject to friction) leaving 1–2 mm diameter superficial erosions. The erosions may remain unchanged until healing occurs or may enlarge and become confluent, forming lesions with a polycyclic or serpiginous outline. The size of the lesions rarely exceeds 2 cm diameter but giant forms occur occasionally. Healing usually occurs in 5 to 10 days but may be delayed by secondary bacterial infection. Moderate, often painful, enlargement of the inguinal lymph glands is found in about 20 per cent of cases.

Lesions may occur anywhere in the anogenital region in both males and females. The mucosal surfaces of the penis and labia are often affected and

496 Electron microphotograph of *Herpes virus hominis type 2* showing internal capsid and surrounding envelope.
497 Early vesicular lesions of penile shaft.

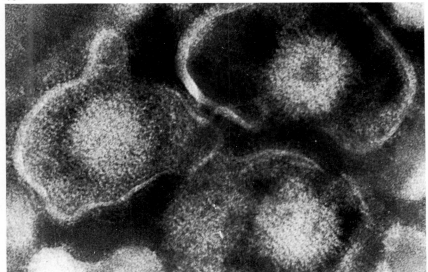

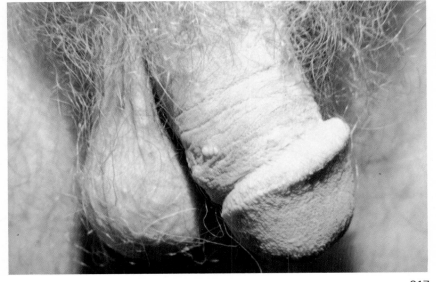

herpetic lesions may be found on the cervix and, rarely, in the urethra. Urethral lesions may cause symptoms of urethritis: the usual finding of other adjacent herpetic lesions makes the diagnosis fairly simple, but in some cases recurrence of herpes can be caused by concomitant urethritis of other aetiology.

First attacks in adults are often very painful and associated with mild constitutional disturbance. Severe dysuria is frequent in females and urinary retention may occur. Recurrences are usually less painful or are irritant only and nearly always occur at the site or sites of initial infection.

Diagnosis

Diagnosis of genital herpes is usually easy, particularly in recurrent attacks but considerable morphological variation occurs: examination to exclude syphilis is essential.

Virus may be grown in tissue culture inoculated from lesions. Characteristic changes due to herpes may be recognised in cervical exfoliative cytology.

Complement-fixation tests may be helpful, especially in first attacks, and serial examinations show change of titre. In recurrent attacks titre changes are unusual.

498 Early papular lesions and erosions of penile shaft.
499 Early papular lesions of penile shaft. The oedema surrounding the group of papules is very characteristic: the whole area is usually irritant.
500 Early vesicular lesions and erosions of perianal region.
501 Multiple superficial sub-preputial erosions. Note irregular morphology.

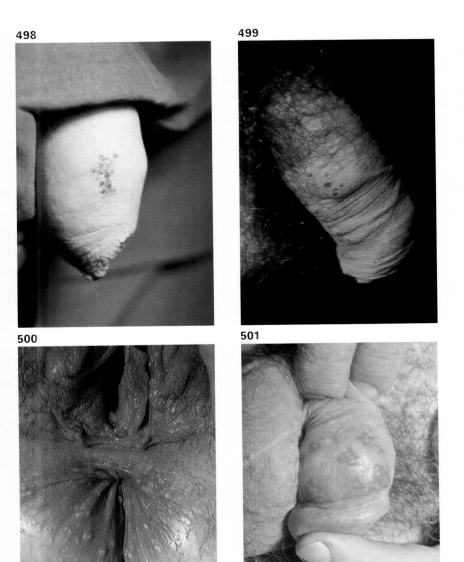

498

499

500

501

502 Multiple tiny erosions of glans penis and prepuce: herpetic balanitis.
503 Solitary large erosion of penile shaft.
504 Typical herpetic erosions of coronal sulcus. Note the adjacent erythema and oedema.
505 Giant herpetic ulcer of penis, with small satellite lesions.
506 Herpetic erosions of prepuce, corona and glans penis.
507 Healing group of early papular lesions of penile shaft and active erosions of prepuce.

502

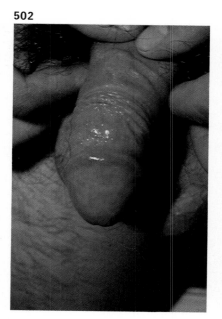

503

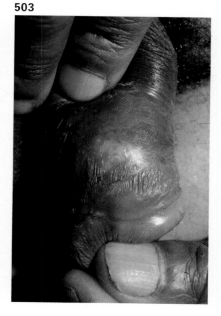

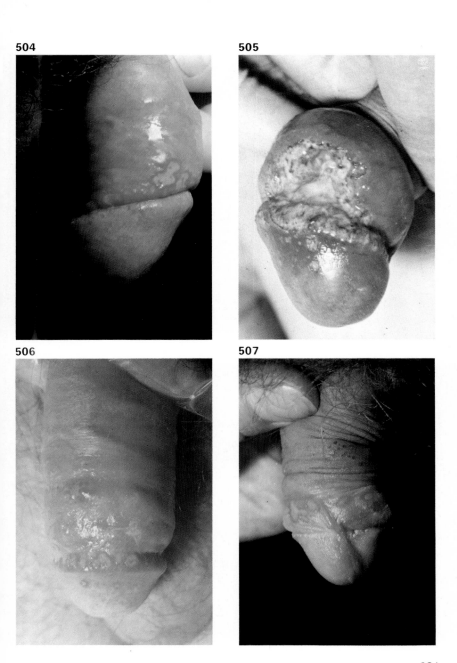

504

505

506

507

508 Atypical herpetic erosions, deeper than normal. The yellow coloration of the margins of the lesions is due to self-applied medicinal powder.
509 Herpetic erosion simulating balanitis.
510 Herpetic ulceration of labia minora. Lesions of this type are often extremely painful.
511 Vulval herpes, showing active lesions and depigmentation at sites of previous lesions.
512 Herpetic erosions of labia, perineum and anus.
513 Extensive secondarily infected herpes of labia.

508

509

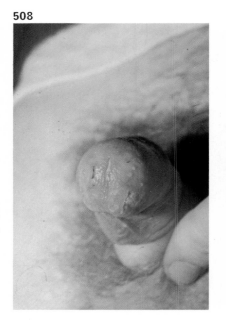
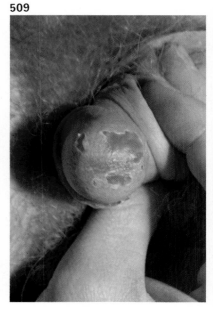

510

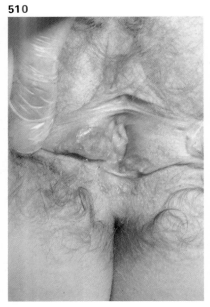

511

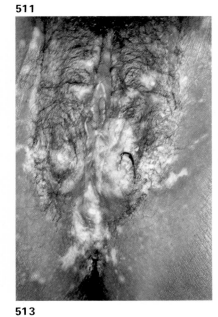

512

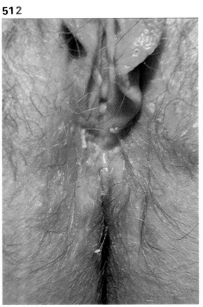

513

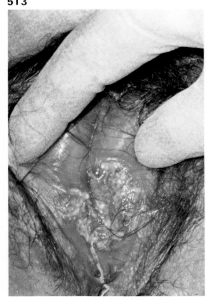

514 Multiple herpetic erosions of vulva.
515 Herpetic vulvitis and lesions of buttock.
516 Close-up of fig. 515 showing perineal and perianal lesions.
517 Healing herpes of cervix.
518 Active herpes of cervix.

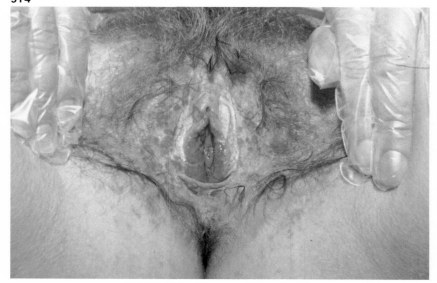

515

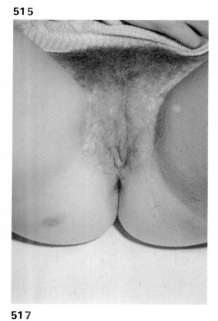

516

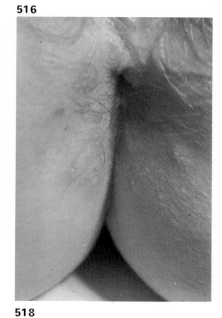

517

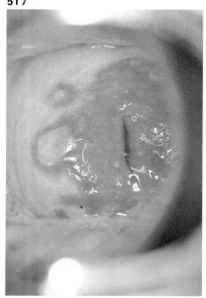

518

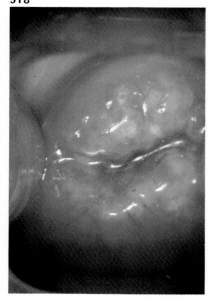

519 Herpes of cervix.
520 Extensive herpes of cervix.
521 Perioral herpes: the patient has concurrent genital herpes.
522 Herpetic vesicles on pharynx: concurrent genital herpes.

519

520

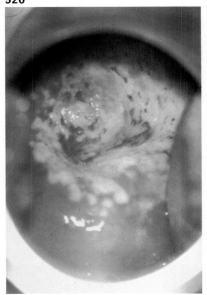

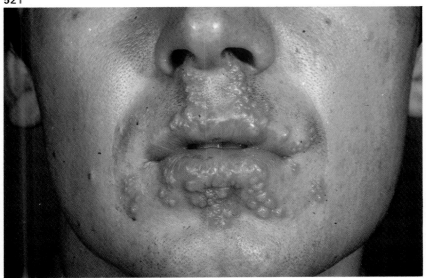

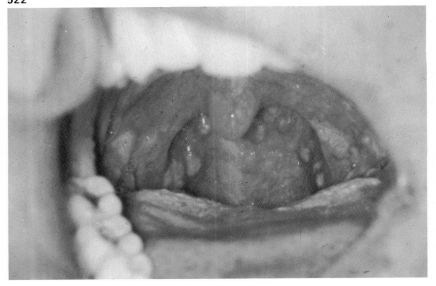

Herpes zoster

Herpes zoster (shingles) occasionally affects the anogenital region. The clinical findings are similar to those in *herpes simplex* infection, but the symptoms are usually more severe and lesions persist for several weeks. The first manifestation of infection is usually localised hyperaesthesia, followed a few days later by the typical vesicular eruption. Symptoms and signs are almost always unilateral: recurrence is rare, although localised hyperaesthesia may persist for some time.

523 Herpes simplex of pubis.
524 Herpes zoster: thigh and pubis
525 Herpes zoster: penis
526 Herpes zoster: buttock and scrotum

Note the unilateral distribution of the lesions.

523

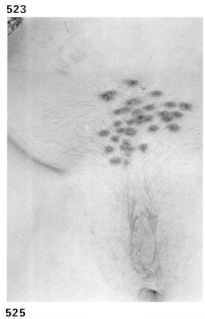

524

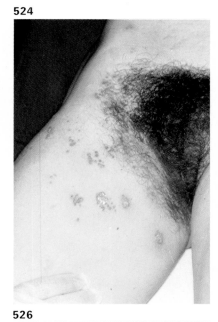

525

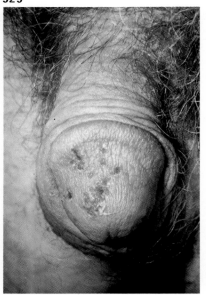

526

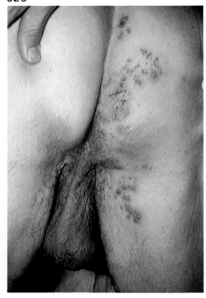

Molluscum contagiosum

This condition is caused by a virus which is transmitted by direct contact, and is a common finding in young people. After an incubation period of two to seven weeks, small, shiny, umbilicated, hemispherical, white papules appear at sites of inoculation. The lesions gradually enlarge, reaching a maximum diameter of 8–10 mm. Molluscs may be found anywhere on the body: in the genital region they are usually seen on the pubis or penis. Lesions are asymptomatic or slightly irritant. Occasionally they may become erythematous from excoriation.

527 Molluscum contagiosum of penile shaft.
528 Molluscum contagiosum of penis.
529 Molluscum contagiosum of mons.
530 Molluscum contagiosum of pubis.

527

528

529

530

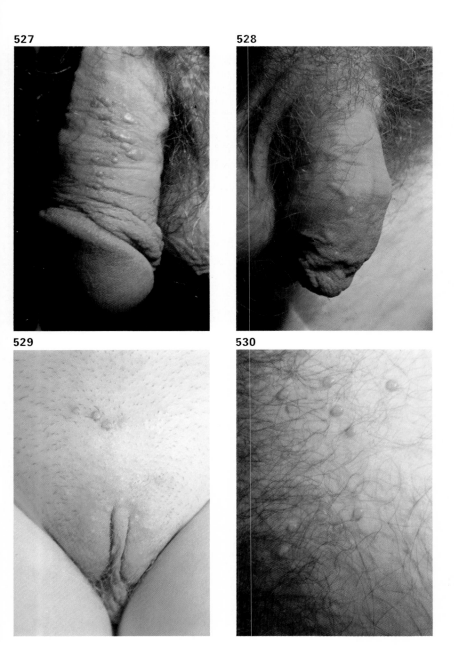

Genital warts

Genital warts (*condylomata acuminata*) are papillomata frequently seen in both male and female patients. They are caused by infection transmitted by sexual contact in the majority of cases: rarely (in one per cent or less) ordinary skin warts (*verruca vulgaris*) are located on the genitalia. A large proportion (60 per cent plus) of the sexual partners of patients with genital warts are found to have or to develop the condition. The epidemiology of wart infections is strikingly similar to the pattern observed in other sexually transmitted diseases.

Cause
The virus responsible for the majority of infections is probably a member of the *papova* group.

Clinical course
Warts are essentially asymptomatic but the appearance, location, size and the presence of associated conditions may cause attendance. Bleeding from the warts may occur: when located in the urethra mucoid discharge is often found. Secondary infection may cause offensive odour, irritation and mild discomfort.

The warts begin as small papillomata, usually multiple. Subsequent growth may produce either filiform, hyperplastic or sessile lesions, depending on the anatomical location. In both sexes warts are found most commonly in areas subject to trauma during sexual activity. Neoplastic change has been reported, but it must be extremely rare.

In *male patients* warts are found slightly more frequently in the uncircumcised. Lesions are found most frequently adjacent to the fraenum, on the glans, in the coronal sulcus and inside the urinary meatus. The shaft of the penis is occasionally involved, and in passive homosexuals anal and perianal warts are often seen. Growth in the urethra and anus occasionally extends proximally for 1–2 cm. The majority of warts seen are filiform or hyperplastic; sessile lesions are found on the glans in circumcised patients and on the shaft of the penis.

In *female patients* warts are seen most often at the fourchette, on the labia (minora and majora), the perineum and in the perianal region. Warts are occasionally found on the vaginal walls and cervix. Hyperplastic lesions are much more common than sessile lesions. Occasionally extremely large masses of warts develop, particularly during pregnancy.

531 Electron microphotograph of wart virus.
532 Rosary of warts.
533 Warts in coronal sulcus: a very typical location.

531

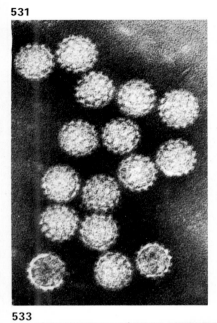

532

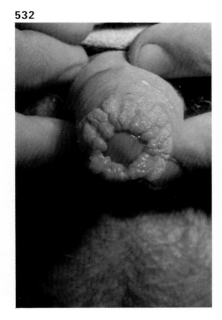

533

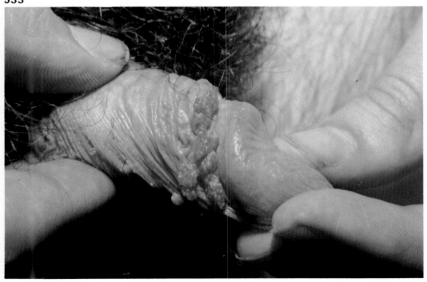

Giant condylomata acuminatum (Buschke–Lowenstein)

This is an unusual and very rare variety of wart infection. The lesion is a single wart with extensive confluent superficial spread and occasional invasion of underlying tissue. It is reported to be more common on the penis.

534 Plane warts of prepuce and glans penis.
535 Meatal and fraenal warts.
536 Gross wart infection.
537 Common warts (verruca vulgaris) on penis.
538 Small perianal warts.
539 Massive perianal warts in male homosexual. Compare with fig. 542.

534

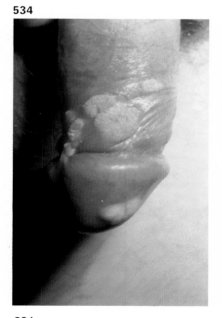

535

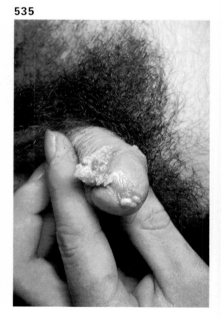

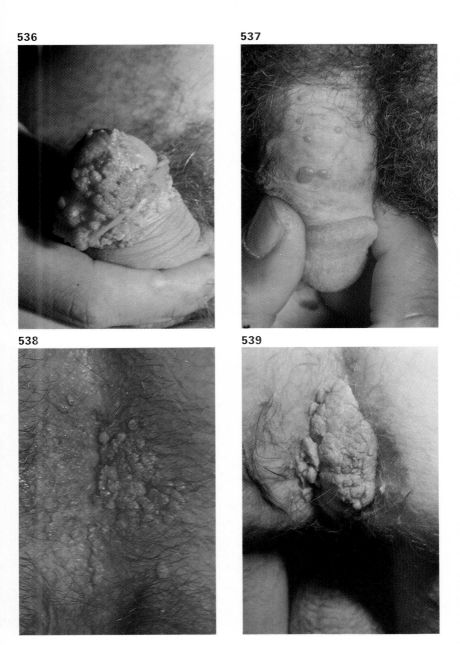

536

537

538

539

540 Labial, perineal and perianal warts.
541 Vulval warts.
542 Massive vulval warts and intertrigo. Compare with fig. 539.
543 Multiple small warts: this type of lesion is often described by the patient as a 'rash'.
544 Wart on cervix. The yellow colour is due to application of podophyllin.

540

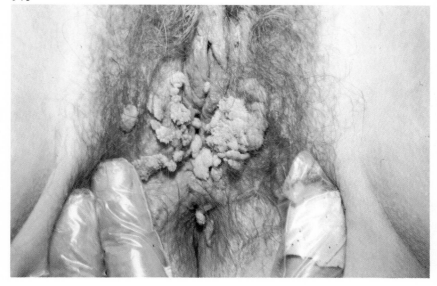

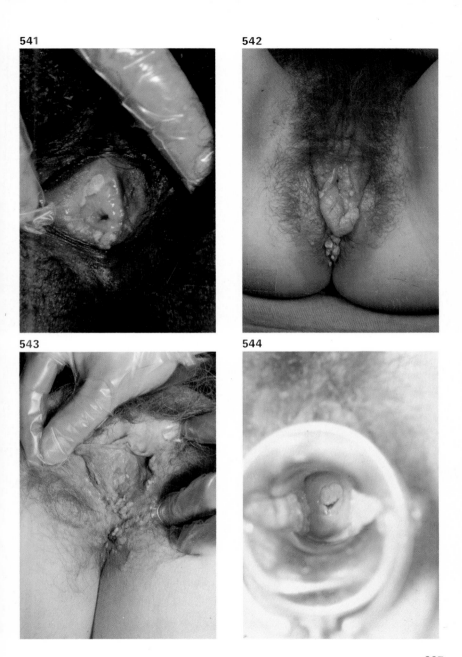

MISCELLANEOUS

Peyronie's disease

The cause of this condition is unknown, but there is an association with the
lesion known as Dupuytren's contracture. Fibrous nodules appear in the
corpora cavernosa. The nodules are usually noticed accidentally but
occasionally complain of deviation of erection (*chordee*) is made. The painless nodules may be single or multiple and vary from pea to grape-size;
occasionally nodules are confluent. The fibrotic masses may enlarge or
regress but the factors which influence change are not known.

Tuberculosis

Cutaneous genital tuberculosis is now an extreme rarity. The diagnosis is
usually established with difficulty by bacteriological or histological methods.
The disease may present as chronic painful ulcerative lesions of the penis
or vulva (*tuberculous chancre*) or as a *'cold abscess'* of the groin. Rarely,
visceral tuberculosis may present as chronic salpingitis, chronic epididymitis
or chronic urinary tract infection. Investigation for *Mycobacterium
tuberculosis* is worthwhile in patients with particularly persistent non-
gonococcal urethritis as a small proportion (one per cent or less) are found
to be due to this organism. Sexual transmission is said to occur.

545 Patient with plaque of Peyronie's disease near base of penis;
curvature of the penis (chordee) is demonstrated.
546 Tuberculous sub-preputial ulcer. Diagnosis was established by biopsy.
547 Tuberculous ulcer of the meatus. The diagnosis was established
bacteriologically.

545

546

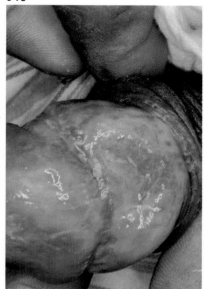

547

Stevens–Johnson syndrome (erythema multiforme)

The exact aetiology of this condition is unknown but in about 50 per cent
of cases there appear to be antecedent factors which include drug
sensitivity and various infections such as herpes, syphilis and lymphogranuloma venereum. Manifestations follow the precipitant (if present) after
an interval of one to three weeks All cases show haemorrhagic bullous
and erosive lesions of mucous membranes which affect the penis,
vulva and conjunctivae. Stomatitis is usually severe and characteristically
causes haemorrhagic crusting of the lips. The genital lesions may simulate
balano-posthitis or vulvo-vaginitis: if the urethra is involved non-gonococcal
urethritis is often found. Conjunctivitis occurs in most cases and may be
complicated by corneal ulceration. Most cases show a skin rash with
characteristic 'target' lesions (cyanotic centrally, erythematous at periphery),
occurring in crops and principally affecting the limbs. Considerable systemic
disturbance and fever are common. Diagnosis is made on clinical grounds;
complete recovery occurs after two to three weeks.

548 Erythema multiforme: typical 'target' lesion on arm.
549 Erythema multiforme: balanitis.
550 Erythema multiforme: typical lesions on thigh and penis.

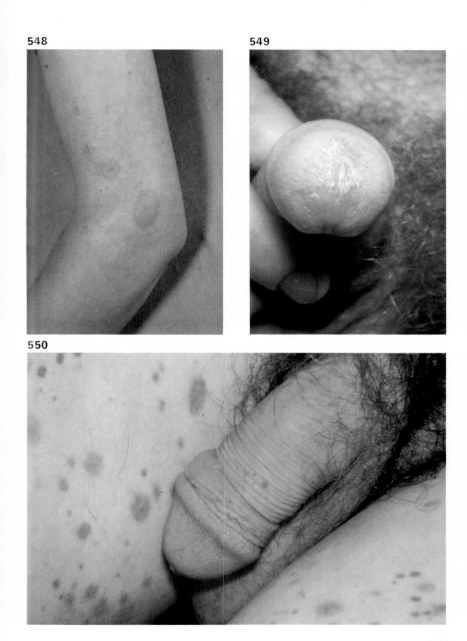

551 Erythema multiforme: stomatitis: note the crusted blood on the lips.
552 Erythema multiforme: lesion inside lip. Compare with mucous patches in secondary syphilis.
553 Erythema multiforme: lesions of tongue.

551

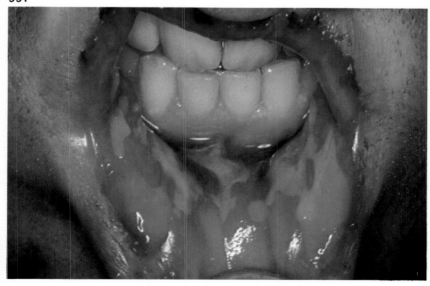

552

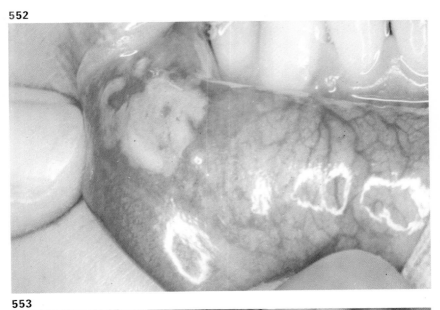

553

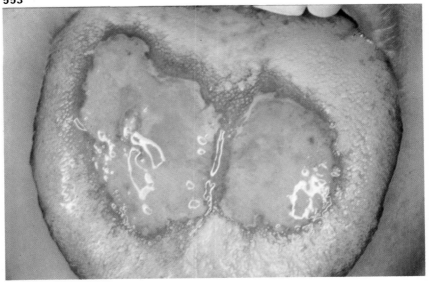

343

Crohn's disease

This condition, also known as *regional ileitis,* is an intestinal disease of unknown aetiology which frequently involves the rectum and anus at an early stage by the development of perianal and perirectal abscesses which commonly form fistulae on the skin adjacent to the rectum. The fistula may appear to be a granulomatous ulcer and, as such, may provoke attendance, particularly in male passive homosexuals. The diagnosis is usually established by the history of intestinal disorder, radiological investigation and by biopsy of the superficial lesion. The ulcer is frequently painless, and may be the sole manifestation present.

Cutaneous amyloidosis

This rare condition of unknown cause has several different clinical patterns: the diagnosis is usually established by biopsy. Localised tumours may occur anywhere on the skin surface and the lesion is usually asymptomatic. Diffuse infiltration may also occur. The disease may be idiopathic or may occur at sites of existing lesions.

Leprosy (Hansen's disease)

The cause of the chronic infectious condition is *Mycobacterium leprae* (Hansen's bacillus). Genital lesions are rare but typical macules, papules, plaques and nodules may occur. The lesions are usually multiple and may be hypopigmented or slightly erythematous: anaesthesia is unusual. Usually other manifestations of leprosy are present.

554 Crohn's disease: perianal ulceration, sometimes the presenting manifestation. Note the similarity to anal primary syphilis, fig. 94.
555 Amyloidosis of vulva.
556 Leprosy of scrotum. Compare with fig. 471, scrotal folliculitis.

554

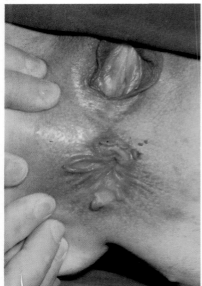

555

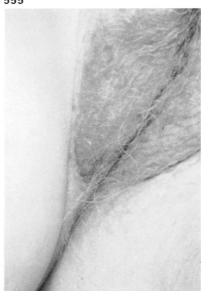

556

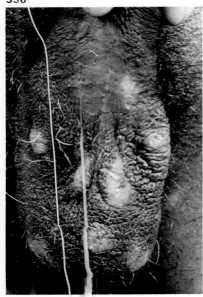

INDEX

*(The references printed in **bold** type are to picture numbers, those in light type are to page numbers.)*